Chinese Space

For several decades now, China has been accelerating its pursuit of greater international prestige and influence. One increasingly prominent facet of that pursuit is the development of an independent space program. This book explains how China is now able to hold such ambitions and analyses how the interaction between technology, politics and economics has influenced the Chinese space program. The book opens by tracing out the earlier development of the space program and identifying the successes and problems that plagued this initial effort. It then focuses upon its development over the past decade and into the future. As China is now able to reach into outer space with its technology and, since 2003, with its humans, the authors analyze how this move from a non-participant status to a state operating at the highest level of space activities has confirmed its potential place as the new economic and military superpower of the twenty-first century. This volume also demonstrates how recent successes mean that China is now confronted by the one issue previously encountered by other space "powers," such as the United States and the former Soviet Union: what is the value of the space program, given its high costs and likelihood of dramatic failure?

This book will be of great interest to students of space studies, Chinese politics, security studies, and international relations in general.

Roger Handberg is Professor of Political Science and Department Chair at the University of Central Florida. His most recent books include *International Space Commerce* (2006), *Reinventing NASA* (2003), and *Seeking New World Vistas: The Militarization of Outer Space* (2000).

Zhen Li is a PhD student in the Department of Political Science at the University of California at Davis.

Series: Space Power and Politics
Series Editors: Everett C. Dolman and John Sheldon, both School of Advanced Air and Space Studies, USAF Air, Maxwell, USA

The *Space Power and Politics* series will provide a forum where space policy and historical issues can be explored and examined in-depth. The series will produce works that examine civil, commercial, and military uses of space and their implications for international politics, strategy, and political economy. This will include works on government and private space programs, technological developments, conflict and cooperation, security issues, and history.

Space Warfare: Strategy, Principles and Policy
John J. Klein

US Hypersonic Research & Development: The Rise and Fall of Dyna-Soar, 1944–1963
Roy F. Houchin II

Chinese Space Policy: A Study in Domestic and International Politics
Roger Handberg and Zhen Li

Chinese Space Policy

A Study in Domestic and International Politics

Roger Handberg and Zhen Li

LONDON AND NEW YORK

First published 2007
by Routledge
2 Park Square, Milton Park, Abingdon, Oxon OX14 4RN

Simultaneously published in the USA and Canada
by Routledge
711 Third Avenue, New York, NY 10017

Routledge is an imprint of the Taylor & Francis Group, an informa business

First issued in paperback 2012

© 2007 Roger Handberg and Zhen Li
Typeset in Times New Roman by Graphicraft Limited, Hong Kong

All rights reserved. No part of this book may be reprinted or reproduced or utilised in any form or by any electronic, mechanical, or other means, now known or hereafter invented, including photocopying and recording, or in any information storage or retrieval system, without permission in writing from the publishers.

British Library Cataloguing in Publication Data
A catalogue record for this book is available from the British Library

Library of Congress Cataloging in Publication Data
Handberg, Roger.
Chinese space policy : a study in domestic and international politics / Roger Handberg and Zhen Li.
    p. cm. – (Space power and politics)
  Includes bibliographical references and index.
  ISBN 0-415-36582-1 (hardback)
  1. Astronautics – Government policy – China. 2. Astronautics – Political aspects. 3. China – Foreign relations – 1976–
I. Li, Zhen, 1973– II. Title.
TL789.8.C55H34 2007
629.40951–dc22
                                            2006020397

ISBN13 978-0-415-36582-6 hardback
ISBN13 978-0-415-64661-1 paperback

Contents

	Preface	vi
1	Overcoming the past, seizing the future	1
2	China as space follower and leader	34
3	First awakenings	57
4	Accelerating the rise of China's space program	84
5	The politics of Chinese human spaceflight	127
6	Assessing China's future in space	151
	Appendix A: Chinese launch vehicles	174
	Appendix B: Chinese strategic missiles	176
	Notes	177
	Selected references	191
	Index	197

Preface

The Chinese space program remains literally a work in progress – changes and advancements come with great rapidity. The long awaited take-off stage has been achieved where China can build on its human and technological capital – both were accumulated at great cost especially for an economically developing state. As will be seen, success was neither immediate nor continuous; their program was not the march of inevitable progress. Setbacks occurred, usually as the result of political interventions into the space program, but especially in the early years as new technologies failed as the Chinese pushed beyond their capabilities. Now, their capabilities have moved into line with their aspirations.

China is climbing two stairways to the heavens. First, and most obvious, China has reached outer space with its machines and now humans. China's anticipated goals in the near future now include taikonauts reaching the lunar surface, a feat only accomplished once before in the late 1960s and early 1970s. The second stairway is China's moving up the hierarchy of space-participating states. China has moved from a non-participant status to a state operating at the highest level of space activities, human spaceflight. Its journey up has increased Chinese prestige and status across the globe, while raising anxieties among some. China always sought to return to what its leaders considered its rightful place as a major power in international politics, and that ambition is coming closer to fruition.

China's successes, however, raise the question of in what direction its space program should move, especially the human spaceflight aspect. That activity is the most difficult and most prestigious but its benefits, at least in the short term, may be overtaken by the heavy costs incurred. Like the United States and the Russian Federation, China now has to be careful about what it wishes to accomplish because, in achieving those goals, the price may become higher than is acceptable in a society with very real needs.

Space activities are open to all states willing to pay the price of admission economically and technically. China becomes a case study in how a state formerly thought unprepared for such activities can now reach for the heavens. Its success points to the fact that space activities properly conceived and executed can be a real asset in the battle to overcome economic underdevelopment and all the assorted issues that go with that status.

Chapter 1
Overcoming the past, seizing the future

Introduction

Present Chinese space policy represents a long tale of struggle both domestically and internationally as a historically great power sought to return to international prominence. China now stands at the pinnacle of the international space prestige hierarchy, alongside Russia and the United States, with the launch of its Shenzhou 5 spacecraft carrying Yang Liwei, a Chinese astronaut or taikonaut. Space programs were conventionally considered a game of the rich country club due to their large resource requirements, but China's rise as a significant space power challenges this perception.

How did a country like China, considered backward both economically and technologically, emerge as an important player in the arena of space technology? What is the motivation of the Chinese government for engaging in its space program over the years? The analysis presented in this book traces the evolution of Chinese space policy and explores the broader historical and political context within which the Chinese space program has developed. This includes the now defunct Cold War, ending in 1991 with the disappearance of the Soviet Union, with its peculiar stresses and issues. In addition, the post Cold War dynamics and all the other major factors that impacted China's domestic struggles in space program development are discussed.

Internationally, in the post Cold War world, China is becoming a more important international player as it fully enters global politics, a journey fraught with setbacks. Domestically, China is transforming from a former socialist state to a hybrid state characterized by a free market economy and a one-party political system. The transition also creates certain internal dynamics, which impact policy makers' perceptions which in the past were often motivated by internal domestic political struggles which both hindered and facilitated the program's development. When supporters of China's space program firmly hold power, resources flow to the space program more readily than otherwise.

The space program itself was never a major independent center of domestic political conflict but was bundled up with larger issues of government science and technology policy along with its ever-present defense concerns. What has changed for China is that its space program has grown in stature and scope to the point now that the national leadership is committed to its success regardless of other factors. All of this national activity must be placed in the larger context of China's historical role in international space policy development.

The international space regime clearly reflects the dominant influence of the original space participants, the Soviet Union and the United States. China represents a clear political challenge to those arrangements along several dimensions, political and developmental. China's influence arises as the result of alliances with others and the improving state of development of its space technologies. Above all, China demands to be treated as an equal partner or competitor to the other two participants – it is unwilling to accept second-tier status. In this area, China has moved from the status of underdeveloped country to that of world-class space participant. For many years, China's aspirations and rhetoric outran its capabilities – something considered no longer true in the twenty-first century.

This study of the Chinese space program is not simply a description of the technologies employed in their space program, but those technical factors will be described when germane to the analysis. Nor is it strictly a detailed blow by blow history of the Chinese program, although the general historical context will be provided. Rather our intention is to place China explicitly within the larger contours of international space policy development, an historical process beginning with the public inception of the space age in October 1957 and even before that event. China, like Japan and India, represents a country aiming to develop its space program as independently as possible.[1] In India's case, that independence was forced on it due to its concomitant nuclear ambitions which isolated India with regard to missile or rocket technologies. Japan's space ambitions have fluctuated over the years, a reflection of its defeat in World War II and the resulting constitutional restrictions placed on military technologies.

China's struggle for national autonomy is one factor in creating a Chinese space program strong enough to stand on its own, and, more importantly, strong enough to cooperate on an equal basis with other states in multinational and international space projects. Cooperative international space activities provide China with entry into a variety of other important contexts: economic, technological and political. In this analysis, the central historical argument is that Chinese space endeavors have often mirrored the earlier space efforts by the now defunct Soviet Union and the United States. Building and operating a successful space program demands that certain technological and physical tasks be completed, such as achieving access to orbit or building viable spacecraft once in orbit. Those steps are not avoidable although states clearly organize their respective space programs in line with their own culture and politics.

China, due to various political and military disputes, found itself building a space program in a manner analogous to the original pioneers. That isolation plus economic and technological weakness meant that China's journey to space was hard and often beset with failure. The early space age saw the United States and the Soviet Union confront similar disappointments and failures but their economic and technical resources were greater, plus the goad of the nuclear arms race kept their attention focused on success. China built its program from a much less secure economic foundation, lengthening the time needed for success. In addition, Chinese political upheavals further disrupted progress. All of this meant that Chinese progress and success were not clear internationally until fairly recently. For years, China was off the radar of most observers.

Since October 15, 2003, China has stood at the pinnacle of the international space prestige hierarchy, alongside Russia and the United States, with the launch of its Shenzhou 5 spacecraft carrying Yang Liwei, a Chinese taikonaut. China and Russia and the United States are the only states to engage in the independent launch and recovery of humans to Earth orbit. All other astronauts, cosmonauts or *taikonauts* (or *yuhangyuans*) who venture to Earth orbit will do so as passengers of these three. The Chinese, like the earlier two programs, will only fly its nationals at first but that will change once its human spaceflight program becomes more established. Individuals from twenty-nine nationalities have flown to outer space, but always as invited or paying guests.[2] This feat of reaching Earth orbit obviously could have been accomplished by other states, but for various political and financial reasons those states have chosen not to do so. In fact, that political reality explains why by 2004 only seven states operated "their own launch vehicles and launch sites."[3] Other states such as Brazil are on the verge of independent success but launch failures and other accidents have crippled their efforts.[4] Technology remains an important hurdle but for advanced states that usually represents a surmountable obstacle if the political and economic desire is present.

Various reasons are given for not pursuing launch to Earth orbit, including the likely costs versus benefits, other international priorities, and various permutations of always-present budgetary tradeoffs within each state's domestic political context. China's success came only after many years of often-solitary struggle and stands unique among those states classified as economically underdeveloped. China remains a state encompassing great contrast between its cutting edge industries and universities in the coastal areas and urban centers and the much more traditional and poorer rural interior areas. This work presents one perspective on that national journey, which began with the dawn of the space age on October 4, 1957 when Sputnik 1 flew to Earth orbit. China's earliest space efforts were largely ignored in the West (the noncommunist world) except for their dual use capabilities. Rockets to orbit can become missiles sent to targets across the globe. The first two space participants, the United States and the former Soviet Union, were especially concerned with this aspect given their political difficulties with China.

One of the more interesting questions for the serious study of space activities is figuring out what factors motivate particular states to engage in human spaceflight. The question becomes an important one because several states or organizations, such as Japan and Europe through the European Space Agency (ESA), have either significantly slowed their earlier human spaceflight efforts or effectively stopped development for the present. The European Hermes spacecraft was stopped in the 1990s as a result of cost questions. That decision does not mean those states are totally disengaged, but their major human spaceflight activities occur through use of other states' launch vehicles, those of the Russians or Americans up to now. China, by virtue of its first human to orbit, now moves higher up the list of prospective space partners. For example, China has become a partner in the European Galileo navigation satellite system, with future plans for other cooperative activities. China and European states had considered cooperative space activities years earlier but Cold War tensions hampered any such efforts. As will be discussed, China's partnership opportunities were long hampered by its technological backwardness. Once that problem was overcome, more advanced states became interested in partnering. Japan, a regional competitor to China, has found any cooperation with China more difficult politically.

One obvious question is why China pushed forward to embrace human spaceflight while other states, clearly initially economically and technologically stronger, have faltered in that quest. One intriguing argument is that only those states that "fear" for their survival have engaged in development of independent human spaceflight. The fear identified here grows out of those states' sense of military confrontation, possibly even encirclement, by other stronger states. This clearly drove the American–Soviet space race; China likewise feared the other two. All three felt threatened in some fashion but China was objectively more threatened because of its obvious military weakness, possessing no nuclear weapons until the late 1960s.

Conversely, China's pursuit and ultimate possession of nuclear weapons fed those fears by the others, the essence of the security dilemma. Each side perceived the other as irrational and especially dangerous once nuclear weapons entered the equation. Nuclear deterrence may work in practice but its realities are always more frightening to those powers considered weaker. Doomsday scenarios are considered a reality when a state feels directly threatened by nuclear weapons. Massive retaliation as an operational concept for US nuclear forces represented a very real threat for Chinese leaders who had already been confronted by President Eisenhower with use of nuclear weapons if the Korean War was not terminated.[5] China's problem was twofold: first, most critically, China lacked nuclear weapons; and second, it had no capacity to attack its potential enemies in either of their heartlands. Deterrence, from the Chinese perspective, was clearly one-way; initially the Chinese were the deterred, although Mao thought otherwise, but he never acted on his belief. Redressing that imbalance became a major goad for pursuing China's nuclear and space programs. The Chinese space program was at first essentially the pursuit of ballistic missiles, and then space launch.

For all three states, building military rockets was their entry into the space age. Neither the Soviets nor the Americans set out to reach Earth orbit as their primary goal. Both states built ballistic missiles of varying ranges – technologies readily convertible to space launch vehicles. Once their nuclear arsenals reached a certain size (a constantly moving target for years) and were robust enough to deliver warheads, the bilateral arms race became too dangerous. Each side needed to demonstrate their capacity to threaten the other without actually heightening the threat level. Other means had to be found for demonstrating one's missile capabilities. The space race fitted into that paradigm and the two states were off and running. Thus, China finds that one benefit of its space program is as a demonstrator of Chinese military technology since most observers assume the military hardware is better than the civil.

For the two space participants, that nuclear fear provided the necessary impetus and urgency that motivated a state to expend the large amount of resources required to place humans in outer space and return them successfully. Firing a rocket with its payload of humans to Earth orbit becomes a surrogate demonstrator or validator of their capacity to launch missiles to the foe's territory. Up to this point in history, human spaceflight has not been militarily or economically relevant. Automated systems efficiently accomplish all the necessary military and commercial tasks that are presently conducted in outer space.[6] The true political value of space does not come from the successful launch of humans but from the political and other implications drawn from that act. China's leadership has understood that reality – a fact which sustained its efforts over the years once a sufficient technological base existed to accomplish the task. Failure is especially debilitating politically.

Despite China's huge population and geographic area, its size alone did not obviously or logically make China the third state to launch humans to Earth orbit after the Soviet Union/Russia and the United States. In fact, one could strongly argue that diverting the human and economic resources necessary for human spaceflight was a decision contrary to China's long-term economic and social interests. For China, as for the United States and the Soviet Union, however, the international prestige along with the technological cachet associated with human spaceflight justifies the added expense. The benefits received are both economic and technological in nature but the political implications are not insignificant. Political implications, however, come in widely different forms. During its early period of development, China's activities were viewed with great alarm by the United States. Now that China has reached a certain level, its participation in global projects has become more acceptable.

One early sign of that political change occurred with China's inclusion in the early discussions on international cooperation in the US-led Vision for Space Exploration.[7] The Vision for Space Exploration, announced by President George W. Bush on January 14, 2004, is a multiyear or decade program aimed at returning humans to the Moon first and eventually sending them to Mars.[8] Chinese participation in the early planning process was a significant albeit indirect ratification of China's heightened status within the space community. China had earlier been excluded from joining the International Space Station program due to US opposition, a situation not changed yet but likely to happen. The George W. Bush administration and China view each other with great suspicion, but space in this limited context can become one bridge between the two states who disagree on many other things.

Other states, especially China's most likely potential enemies internationally, are not awed by the October 2003 flight (they are already long-term space participants at some level) but the new reality gives them pause in considering possible future acts of aggression against China, as earlier between the United States and the Soviet Union. Launching a payload to orbit is no longer solely the marker of a technologically sophisticated state but sending and returning a human remains singularly unique (only three states have done so). Regionally, the political benefits at this time may be more impressive since none of the other East Asian or South Asian states have demonstrated an equivalent technological capability, although all engage in some space-based activities. India and Japan are long-term Chinese competitors but their governments remain reluctant to commit significantly more resources. The reasons for their hesitancy are different, reflecting national resource differences and political needs, although China's assertiveness will likely generate a response. A space race along the Western Pacific Rim and South Asia becomes a distinct possibility, depending on how China exploits its very public and successful human spaceflight effort.[9] Japan is becoming particularly sensitive to China's policies, especially those with military implications, although North Korea has had a more immediate impact on Japan's space activities.

Assessing the situation, a first cut

As we mentioned earlier, present Chinese space policy represents a long tale of struggle both domestically and internationally as a historically great power sought to return itself to international prominence. Policies regarding Chinese space activities have evolved in the context of a political system characterized by both great upheaval and opportunity. The analysis presented here traces the evolution of Chinese space policy from the perspective of its efforts at establishing China as a major global political-military player, including others' reactions to China's efforts. Chinese policy generally responded to what China perceived as hostile actions and intentions by the United States first and later the Soviet Union.

This particular analytic perspective in a very important critical sense places China on a par with the Soviet Union and the United States during the earliest days of the space age. China's historical role during that original period was that of observer, while internally the political and technical foundations of Chinese space activities were being slowly laid in place. The path traced by China's space program was neither smooth nor automatic or inevitable since the escalating political and social demands from the larger Chinese society for more rapid economic development did not always accommodate such esoteric activities at first. The latter's importance grew much clearer in more recent years when China increasingly emerged from its political and economic isolation. China's earlier political and economic isolation was a combination of self-imposed actions and decisions by others. The United States systematically worked to totally isolate China, while the Soviet Union grew more hostile in the 1960s era. The Europeans and Japanese were generally more open to China economically but were constrained by US hostility to expanding other relationships. Over time, tensions eased but a general wariness characterized each state's relations with the other.

Even more disruptive, not just for the Chinese space program specifically but for the entire society, were the political upheavals associated with the transition of power from Mao Zedong and his closest associates to other members of the Communist Party ruling elite. Authoritarian political systems especially confront great uncertainty during transitions of power. China entered its fourth generation of political leadership with Jiang Zemin's retirement. Those disruptions, especially the internal political conflicts during the Cultural Revolution, slowed progress after a very auspicious start in the early 1960s. Some work proceeded forward but critical momentum was lost, something only slowly recovered. In a more dramatic fashion, China's experience mirrors, albeit with critical differences based on culture and economics, similar events that occurred in the Soviet Union and the United States, though not as dramatically. All government-sponsored space programs are continually subject to the shifting winds of domestic political decisions because national priorities constantly change in response to ongoing events. For all states, democratic or otherwise, changing national economic conditions mean changes in national space budgets. Especially for developing states, such changes occur because their list of priorities always outruns their available resources. Their economic margins remain thin, making caution the operative approach. China lacked the wherewithal to engage in massive projects in the same way as the Soviet Union and the United States did for over a decade, ending with the Eagle's landing on the Moon in July 1969.[10] Those budget orgies eventually ended when political priorities shifted – the US decline was the more abrupt of the two programs post-Apollo. The Soviet decline was slower but in the end more final with the collapse of the Soviet Union in 1991. In fact, Soviet budget support for their space activities had been so strong for so long that the Russian program could in effect live off that earlier investment for several years after the collapse.

Placing China in the same general historical context as the original space participants in terms of its motivations for pursuing human spaceflight represents a different perspective from that often attributed to China. Since the defeat of the Chinese Nationalists in 1949, who ultimately fled to Taiwan (formerly Formosa), and the seizure of power on the mainland by the Communist Party, China has often been characterized as a deviant state not subject to or participant in the normal rules and expectations of the international system. That characterization was especially true in American analyses of Chinese actions and policies over the years. The Korean War in 1950 was the catalyst for the intense hostility between the two. The United States intervened in June 1950 as leader of a United Nations (UN) coalition in support of the Republic of Korea (the South) when the Democratic People's Republic of Korea (the North) invaded. American intervention reversed the flow of the fighting by driving the North Koreans toward the Yalu River bordering China and Korea. China's intervention in December 1950 surprised UN forces, driving them back and eventually creating a stalemate which was formalized in a military truce in 1953 (a truce still in effect).

The American Department of Defense (DOD) in its most recent National Strategy characterizes China as the next major state threat on the horizon for the United States.[11] In the 2001 Quadrennial Defense Review Report (released on September 30, 2001), the American Department of Defense (DOD) characterized China as a major state threat: "The possibility exists that a military competitor with a formidable resource base will emerge in the region (Asia). The East Asia littoral – running from the Bay of Bengal to the Sea of Japan – represents a particularly challenging area."[12] According to the *Financial Times*, the 2006 Quadrennial Defense Review Report "will take a more pessimistic view of the challenge posed by an emerging Chinese superpower than the 2001 overview."[13] The ongoing war on terror does not involve at this point an identifiable state threat (Iraq being a battlefield), just a list of the usual suspects including Iran and North Korea. This characterization of China as a potential threat tracks US foreign policy attitudes since the Korean War stalemate and the Chinese intervention in December 1950. US massive retaliation as a political-military doctrine arose after the Korean War and the French Indochina debacle. Nuclear weapons were to substitute for manpower, an equalizer of sorts. Those attitudes have waxed and waned in intensity over the years but the essential bilateral relationship has long been characterized as one of suspicion and tension between China and the United States. The opportunities for misunderstandings and incidents become myriad in such a situation.

A uniquely American perspective has arisen that treats China as the other example of exceptionalism operating in the international system besides the United States itself. Both states, it is argued, operate as if the existing international rules do not apply to their actions and policies because both states operate from a sense of historical uniqueness (clearly different in terms of origin) that excuses their behavior (at least in their eyes). Other states, especially the Europeans and the Japanese, are obviously not as enamored with this self-serving American perspective but the most immediate diplomatic result was that China from 1949 onward until the early 1970s was aggressively isolated by the United States from any significant engagement in most of the existing international regime. The Europeans engaged in more formal contacts such as diplomatic relations with China but remained heavily constrained by their security dependence on the United States during the Cold War. China was far away from Europe and poor while the Soviet Union and its Warsaw Pact allies were near at hand and threatening. Chinese space technology could in principle have moved more quickly but early questions regarding establishing and maintaining Communist Party internal political control took precedence, a not unusual judgment for newly established regimes that feel threatened.

Japan's relationship with China has been more complicated, given events during and after World War II. Japan's military linkages to the United States and its nuclear shield severely restricted any interactions. Those issues have not stopped Japanese investment in China but a certain political tension persists. Controversies over civilian massacres by the Japanese during the war, for example, have periodically arisen.

China's diplomatic isolation was further accentuated by their early 1960s split from the Soviet Union, with the result that China was in a situation of potentially great external threat from two sources, the Soviet Union and the United States. China's political isolation was never totally complete but its gradual collapse signaled changes in the international situation, which began in the 1960s and picked up steam by the 1970s. President Richard Nixon and Chairman Mao Zedong met in Beijing in 1972, one symbol of those changing relationships. China, for example, at first in the 1950s, aligned itself with the nonaligned movement, an uneasy fit at times but one useful for accessing the larger international system since China was excluded through American efforts from the United Nations and its related organizations. The Republic of China, formerly Formosa, now

Taiwan, occupied the China seat on the UN Security Council. China's space program has reflected both directly and indirectly all these events and the intense stresses placed upon China as it embarked on the process of national reconstruction. One major domestic political goal became pursuing an international presence more commensurate with China's historical role in the development of a major portion of Asia. Space activities are one facet of that national effort. China historically was the regional superpower in East Asia; its cultural legacy still permeates all the states in the East Asia region.

Politics, economics and technology

Essentially, the Chinese space program is the product of the continuing interactions between politics, economics and technology in approximately that order of importance.[14] Politics as an analytic concept here encompasses the domestic and international politics of China; both facets have interacted in terms of influencing development of national space activities. Space activities are peculiarly driven by politics because vast sums of money and other scarce societal resources must be committed to the space effort with absolutely no guarantee of quick technical success. Success in eventually reaching Earth orbit was not really in doubt by 1940; how quickly useful space activities could be identified and accomplished remained more uncertain. Usefulness can come in several forms, military and economic development being the most prominent while international prestige often tips the government's decision to proceed or not.

For China, as with other space-faring states, space activities grew out of conscious decisions by the political leadership to pursue such exotic endeavors. Space activities were exotic in that the immediate benefits were a combination of technological and intangible – economic returns were more distant. Therefore, space programs usually begin with development of ballistic missiles, flight technologies with a clear military purpose. Obtaining funding for military security uses is generally easier once the national military reaches a certain level of technical sophistication. The technology has to improve their efficiency and lethality. Therefore, flying to Earth orbit is built upon the foundation of Chinese ballistic missile development.

Missiles and rockets are simply the opposite sides of the equation in that with some modifications both missions can be accomplished using the same launch vehicle. Once a certain level of technology is achieved, military missiles and civil rockets often split off on separate paths since the military demands faster responsiveness, which means solid-fueled boosters, while the civil needs greater control over launch, meaning liquid-fueled boosters.

Missiles are presumed to carry warheads to targets at some distance (short range to intercontinental) while the same rockets lift payloads of different masses and sizes to varying heights ranging from the suborbital to the orbital arc and beyond. The orbital arc lies approximately 22,500 miles or approximately 36,000 km from the earth's surface, a location from which the orbiting spacecraft or satellite does not appear to move relative to the ground. The satellite is traveling as fast as the Earth's rotation. Since the mid-1960s, this location has been considered the optimal location for commercial communications satellites (comsats).[15] The other orbital locations are low earth orbit (LEO), which ranges from 300 km to 1,500 km, while middle altitude orbit (MAO) is between 10,000 km and 35,000 km. Other more exotic orbits include: intermediate circular orbits (ICO) – 10,000 km or higher; highly inclined elliptical orbit (HEO) – out to 40,000 km; and the Molniya orbit (moving from 400 km above the South Pole to 40,000 km above the North Pole during each orbit). In addition, orbits can be polar, crossing both poles during orbit. Which orbit is employed depends on the purpose of the payload.

This interchangeability of the same rocket technology represents the essence of dual-use technology. At a certain point, as we mentioned, the military and civil split, with the military pursuing rapid or launch on demand capabilities, which means solid-fueled boosters, while the civil wants more control over launch, which moves them to employ liquid-fueled boosters. For a state such as China, which is overcoming severe technological and economic deficits, dual-use technology effectively maximizes its returns, both politically and economically although the original purposes were clearly military.[16] Chinese launchers are at their core liquid-fueled with solid boosters attached for generating greater lift. This configuration tracks the original development of rocket technologies by Russia and the United States. For China, this interchangeability becomes even more critical given their severe resource constraints. Economic realities are never distant when Chinese leaders make policy in any technology area, including space activities.

Other studies

Analyzing China's space policy, however, has proven difficult for reasons of access to accurate information, but it has becoming increasingly fruitful in more recent years. Earlier studies generally pursued a China-unique perspective or a heavily technology-hardware focused view. Joan Johnson-Freese's 1998 analysis of Chinese space policy explicitly focused upon the cultural uniqueness of China.[17] A major part of her argument was that explaining Chinese space policy became difficult at times due to the unique cultural national parameters permeating its policy process. All decisions regarding policy emerge from within the closed elite running the state and by extension the Chinese space program. The internal decisional processes and the factors impacting those decisions have long been fairly opaque in order to preserve political flexibility and deflect any domestic criticism. The title of Johnson-Freese's work, *A Mystery Within a Maze*, nicely illustrates the central argument.

Here, the argument is that Chinese cultural and political norms are truly important but other realities are as important, if not more so, if technical success is to be achieved. In fact, that same argument can be extended to every national space program when viewed by outside observers. For the Europeans and Japanese, US space policy often appears inscrutable as abrupt changes occur in the US NASA programs driven by US domestic policy needs. The International Space Station program, for example, has been constantly battered by continual changes and redesigns dictated by unilateral US decisions.

Brian Harvey, in contrast, describes the Chinese space program largely in technological terms although those developments were generally cast against a general background of ongoing political events.[18] Those political events were mostly alluded to rather than fully developed since they were clearly not the major thrust of his analysis. Early delays and disappointments in technology development become important given the Chinese space program's recent growth. The abrupt severing of formerly close political relations with the Soviet Union for a time was a hard blow given the Soviet proclivity to tightly control whatever information and technology was provided to China. Despite these restrictions, however, Chinese technical personnel had learned enough for them to be able to continue pushing forward. Also severely damaging were the economic shortfalls and disruptions that occurred as the result of political choices by the leadership in Beijing.

With the end of the Cultural Revolution in 1976, a more normal developmental process ensued from that point. Harvey's second book on China (effectively a second edition of his first) highlighted more of the political side but the primary focus still remained on the technical aspects which were more knowable, although even there successive "surprises" arose forcing changes in first interpretations of Chinese launches and programs.[19]

The Chinese space program was analyzed by Mark Stokes as part of the Chinese military-industrial complex.[20] Chinese space activities, according to Stokes, are an important component in the Chinese military modernization effort, which aims to strike at the enemy's C^4I system (command, control, communication, computers and intelligence) as a shortcut to winning a war. Hence Stokes focuses on the development of a military-related space program, such as missiles, communication satellites, and precision weapons. However, the military-related space program has not been the sum total of the Chinese space program; rather space program activities have stressed civilian use, which became the primary concern in most years of the 1980s. We should note that the pattern of the Chinese space program development is not static over decades; rather it is more dynamic, as any growing space program must be or stagnate into irrelevance. For example, the motivation for economic development did not exist before 1976, so the space program's goal then was strengthening the military forces and enhancing national prestige, while in most years in the 1980s, military use yielded to civilian use. Military security and national prestige enhancement are two rationales in determining the development of space programs, but politics, particularly the leader's political preference, can be a decisive factor, which was particularly true in the Mao Zedong era. Clearly Stokes ignored factors other than military use since that was his entry point and focus. The game for the Chinese leadership changed over the years once they gained the security of having some form of nuclear force.

The argument being advanced here is that the Chinese space program is best understood as one analogous to those developed earlier by the original space participants, the Soviet Union and the United States. This perspective also suggests that many of the same political concerns and issues, which drove those states in their original pursuit of space activities, drive China. Such a perspective does not deny that there exist idiosyncratic aspects to the domestic Chinese policy process compared to others, but economic and technological fundamentals obviously impose physical and fiscal limits upon their options. In fact, one could argue that China's failures at different points in time illustrate the continuing reality of those limits. There exists no nationally unique route to the stars; the laws of physics still rule. There are only unique national programs reflecting domestic conditions and their general role in the international system.

The other factors that heavily impact China's space efforts are the changing and improving state of its economy and the level of technological development. Until those limitations, especially the latter, were largely overcome, although not completely solved, China progressed slowly in its pursuit of space activities. That success, however, is what makes China an interesting prototype for the future, in a manner similar to India. Both states are demonstrating their independent ability to marshal significant national resources in order to accomplish technological feats previously thought restricted only to the developed world. For each state, their space program becomes a public declaration of their improving technological and economics systems. Both states presently stand somewhere between the first and third worlds, struggling to overcome basic economic resource issues while simultaneously pursuing cutting edge technologies.[21]

China's vast population puts enormous pressures upon its national leadership to quickly improve the national economy. Space technologies, properly utilized, can help in that process, especially employing the simple applications such as remote sensing and communications. The vast distances of China, for example, mean "hardwiring" the country becomes a very long-term project – satellite-based communications, however, can help bridge the gap between the present and the future. Remote villages and regions can be brought into the national society, an attractive thought for political leaders, but the larger benefit can be ultimately fostering education and economic growth. Remote sensing includes weather forecasts and disaster assessment and assistance. For example, severe floods repeatedly devastate whole regions of China. Emergency response is significantly improved when the disaster relief teams have relevant information quickly for planning their response. All of these space activities provide direct and indirect economic benefits for China, justifying the expenditures even if the products are not sold internationally. India and China both see space activities as enhancing their rate of economic development.[22]

There arises a synergism between space activities and economic development. As a state focuses on serious space applications aimed at fostering economic development and some success is achieved, indirectly or directly, further resources become available to support further technology upgrades. The process is neither swift nor immediate, which discourages the most economically challenged states. The demands are so pressing that delay is unacceptable. The difficulty comes in making the first investment – there are no excess funds except those generated by reductions in other critical programs. Authoritarian governments in principle can make such authorizations more readily than more democratic regimes. For China, making those first investments was problematic because their technology base was so weak that nothing effective could be done at first. Creating the human capital along with the concomitant technical support became the first stage in the process. Political rhetoric and ambitions at first far outran their capabilities. That is a major part of China's story with regard to its space program. China has grown into its rhetoric, as evidenced by its recent performance, as will be discussed.

Organization of the work

Analyzing Chinese space policy becomes an interesting task given the complexity of the topic and the various directions possible to pursue. In this first chapter, the focus is upon generally describing the interaction of political, economic and technological factors as refracted through the prism of Chinese politics and culture and the evolving international context. Chinese space policy is placed in comparative context using the first two space participants as earlier role models and prototypes of what can happen. China picks its own path but recognition of what has already happened becomes important in assessing its efforts. In Chapter 2, the larger international context within which China has developed its space program is laid out. That larger context includes the Cold War with its peculiar stresses and issues, all of which was played out against the context of China's domestic struggles to develop economically and otherwise. The international space regime reflected the overwhelming influence of the first two space participants, the Americans and the Soviets, and China almost by definition presents a serious challenge to those established arrangements. China's role by default becomes that of challenger to the status quo.

Brian Harvey is right in conjecturing that Mao wanted to put this program with its high strategic significance in Shanghai because Shanghai was his political base.[28] Mao decided to waste huge social resources to shift some space programs to Shanghai after 1966 due to his own political considerations. When the first Shiju satellite was constructed at the end of 1969, Shanghai had become Mao and his wife's political base. Shanghai's position was further enhanced after Lin's death in September 1971, when the Beijing team was estranged from Mao due to Lin's avid support.

Poor capital investment and insufficient S&T investment resulted in a series of failures in the short-lived Jishu satellite series. This project started in January 1970 and was run by the Shanghai Bureau. Little information on these three satellites was available, not because they were for military use, but because they mainly served as propaganda tools with little other value. R&D on the Jishu series was quite possibly repetition of the work the Beijing team had completed years before, but the construction of the first Jishu series took the Shanghai team five years and all the Jishu satellites functioned poorly. The first Jishu satellite was launched after series of failures in July 1975; it burnt up in the atmosphere after fifty days. The second satellite was launched in December 1975 and its orbit decayed after only forty-two days. The third one was launched in August 1976; it survived in orbit for 817 days. All these three satellites were given great political significance by the government headed by Mao in his political campaigns against Lin Biao and later Deng Xiaoping. Even the slightest technological information, such as orbit parameters, was not given, and the West never picked up their signals. Thus whether they did fly as the government then claimed is doubtful.

Although space programs can be political leverage regardless of regime type, it is clear that under the totalitarian regime, the paramount leader was subject to fewer constraints compared to those of other political regimes, and consequently the policy vicissitudes increased. Another subtext runs through the development of the Chinese space program in all its facets. It has been alluded to earlier but now needs to be addressed explicitly: the security dilemma underlying its relationships with other states regionally and globally, in the case of Russia, formerly the Soviet Union, and the United States. A security dilemma as identified by Thomas J. Christensen is when

> mistrust between two or more potential adversaries can lead each side to take precautionary and defensively motivated measures that are perceived as offensive threats. This can lead to countermeasures in kind, thus ratcheting up regional tensions, reducing security, and creating self-fulfilling prophecies about the danger of one's security environment.[29]

Christensen focused explicitly on East Asia with concerns about China's reactions to changes in Japanese military policy. The Chinese space program from the perspectives of the Soviet Union/Russia and especially the United States has been perceived with great suspicion at different times. Evidence of that can be seen in the original US ballistic missile defense (BMD) deployment initiated by Secretary of Defense Robert S. McNamara in the fall of 1967 – the program was defined as an anti-Chinese defensive system. The Anti-Ballistic Missile (ABM) Treaty of 1972 nullified that deployment for nearly three decades.[30]

The suspicions held by the United States were further fed by any space launch capability on the part of the Chinese government. Chinese success in reaching orbit in 1970 only intensified those concerns. For China, however, reaching orbit was only one visible manifestation of its determination to secure its national security against all challengers whether regional or global. The military necessity for acquiring effective intercontinental ballistic missiles (ICBMs) was overwhelming from the Chinese perspective. In the traditional security dilemma literature, an arms race should result, but in this case, China was at first too weak technologically and economically to engage in such competition. Instead, its strategy in the short term became one of acquiring the necessary military technologies to deter any American aggressive acts, although not sufficient to force US withdrawal from the regions, especially Taiwan.

For the leadership of both countries, the policy became one of cautious engagement albeit buttressed by suspicions of the other's motives. China needed external resources and markets to fuel its economy – a fact which dampened much of the overt conflict. Flare-ups occur whenever one side perceives the other as becoming threatening. For China, the recurring controversies over Taiwan proved continued American unwillingness to accept China's view of its borders. Within the United States, as will be discussed, concerns arose over the Chinese space program as an economic competitor in terms of pricing – the argument being that the Chinese program was a non-market driven competitor which could dramatically undercut prices, driving others out of the marketplace. The larger and more ultimately disruptive conflict came over allegations that the Chinese were using investigations of launch failures as a cover by which to acquire sensitive or restricted information in order to improve their space products but also their missile forces. Both of these conflicts will be discussed later but here their importance is as indicators of mutual suspicion and distrust. Both interpreted the other's actions as malign in intent.

Further fueling Chinese suspicion is the return of BMD to an active deployed configuration. The justification is to counter possible missile attacks by "rogue states," in this case, North Korea. The other rogue state according to the United States is Iran, whose space efforts are seen as a "Trojan Horse" by which missiles with greater ranges can be developed under the guise of space launch.[31] But, for the Chinese, the BMD system in Alaska with possible expansion to California signals a deliberate attempt to disarm China with its small nuclear missile forces.

Overcoming the past, seizing the future   21

The security dilemma explains why suspicions intensify but provides no immediate answer to how one breaks the spiral. Rather, the Chinese space program clearly retains its dual nature of having military implications. This fact leads the leadership to emphasize different aspects at different times – the military and nonmilitary are separating because the latter is growing in economic and political importance. Through space activities, China is becoming even more a part of the world community.

Economics and economic competitiveness

Economics enters the political debate in several ways with regard to space activities and whether they should be pursued or not. First, economics in terms of available national resources, human, technological, and capital investment, dictates how fast states can pursue space activities. Poor states have only limited human resources in the form of trained engineers and technicians. For a less developed country, such allocation decisions are clearly not costless but reflect very hard thinking about the relative value of the end product. There are no easy choices in a world of extremely limited resources.

Second, the political difficulty is that the investment demanded is a continuous one rather than a single or short-term allocation. In fact, the surplus resources or the slack resources in the economic system get diverted to this single facet instead of others. Because technology development can be slow and erratic with multiple testing failures, the comparative cost becomes a real problem. States seriously interested in establishing a national space presence intuitively understand that the investment demanded is large, even though in the early stages the amounts are comparatively small. Governments find pursuit of serious space activities with regard to costs is much like the proverbial camel's nose under the tent flap. The process only gets more expensive as the technologies grow more efficient and sophisticated. Some states initiate space activities but falter under the financial burden, as the United Kingdom did in the late 1960s.

For a long time, the argument was put forth that command economies such as the former communist states could easily muster the requisite resources because public opinion, however expressed, was not a factor. In the short term, that is likely true given the coercive nature of the situation, but the dislocations and inefficiencies introduced eventually damage the entire economy. The Soviet Union, for example, produced an excellent and overbuilt space apparatus – one whose redundancies added to the society's economic burdens. By the end, the entire economic and political system collapsed under its own weight due to escalating inefficiencies. China's command economy has facilitated its efforts to build a space program but its economic margins are thinner, slowing progress as needed resources are often diverted to other more pressing needs. Programs get stretched out in time due to resource issues – this happens in developed and developing states but the former are more likely to redress the issue later.

Even for economically developed countries, space activities must compete with other important demands upon limited government budgets. For the Soviet Union and the United States, space activities initially were completely subsidized, especially in the absolutely critical launch sector, by national military budgets. That government subsidy continues in both cases. The only major innovations in the US launch program after the official end of the Cold War (1991) occurred in its Evolved Expendable Launch Vehicle (EELV) program, which was paid for by the US Air Force.[32] The program involved building larger and more efficient models of the earlier Delta and Atlas rockets. The Russian government, the major successor to the Soviet Union regarding space activities, continues its support of rocket technology development. What is interesting is that, in both cases, their launch industries are now supposed to operate commercially. That scenario implies earning revenues and possibly profits which are reinvested into their launcher inventory.

Early on, China's launch program obviously benefited from a similar symbiotic relationship of military and economic pursuits. The military sector for reasons of national security, as always, is at least partially removed from the push and pull of domestic political priorities. The key is the qualification "partially removed" since all government programs regardless of regime type encounter conflicting pressures, especially if their budget is rising seriously; pursuing space activities means budgets continually rise unless the program is able to off-lay the costs to some other sector, possibly in the West the commercial sector. China's program has not reached that developmental plateau yet, but their experience tracks the history of other major space programs in terms of costs and investment requirements.

As space activities become more routine, significant adjustments based on competing political priorities begin to occur – those budget adjustments often lead to funding cuts either in the aggregate or regarding the rate of budget growth. This normalization of the national space effort is good because budgets are secure, but they also become less likely to experience great growth again. Both the Americans and the Russians have reached that point where enormous year-to-year budget growth becomes extremely unlikely. For example, President Bush announced the previously mentioned new Vision for Space Exploration in January 2004. Unlike the earlier Kennedy 1961 Apollo program announcement, no large budget increase (5 percent or more annually) is projected for the first five years of the program. Even that modest goal is proving more problematic than one would expect. How future increases for the Vision will be funded after President Bush leaves office in 2009 remains unclear. NASA will not disappear but its budget is now considered incremental in nature by the US Congress. China has not reached that stage yet.

In the case of China, however, knowing their total space budget becomes extremely difficult. The information is considered politically and militarily sensitive. In addition, the amounts expended vary depending on the exchange rate for the Chinese yuan both officially and unofficially. China normally deals with other states in US dollars or Euros. Therefore, when budgets are examined, the resulting cost projections vary dramatically depending upon the exchange rate. Regardless, comparatively speaking, China's space budgets are rising as it moves forward. However, success breeds resources, especially given the presumed domestic and international benefits the Chinese leadership currently sees resulting from their space program. The implications of this are considered in Chapters 5 and 6 where we focus on human spaceflight and then the future.

Additionally, space activities are now becoming more international in scope especially regarding commercial applications. That means individual national space policy decisions do not occur in a vacuum. The barriers of the past are much reduced although not totally gone. For example, the global launch market in the late 1990s bottomed out as demand drastically fell. A series of large comsat flotillas was delayed or cancelled, such as Telesdic with its projected 840 or 288 comsats in LEO. The overall situation was aggravated by the fact that at the same time a number of new or upgraded launchers entered the market. The result was a quickly saturated global launch market from whose adverse economic effects China was not exempt. This interconnection of domestic and global means Chinese choices may prove ineffective or unsuccessful because of other states' choices. For a national leadership extremely suspicious of those other states' motives, this pattern is distressing but cannot be helped.

Space technologies remain expensive – why that remains so is somewhat of a mystery since other equally complicated technologies lack equivalent high price tags, at least over time after their initial introduction into operation. Costs for space launch, even controlling for inflation, appear inelastic, not declining as production and other efficiencies supposedly occur. The normal cycle for new transportation systems is for costs to decline over time as the technology is further refined and improved while produced in ever larger numbers, an economy of scale situation in part. Regardless, all state leaderships must make choices as to whether those expenses are bearable. Achieving success in technology development does not automatically translate into any concrete societal economic benefits aside from jobs in the short term, especially if the government and society are not organized to grasp the new benefits generated by their investments. Both the United States and the Soviet Union expended enormous sums to pursue space activities such as remote sensing – a sector whose full economic potential was not realized at first due to national security restrictions and ideological blinders. As those limitations were slowly removed after 1991, benefits began to accrue. The security restrictions on remote sensing flowed from the reality that imagery produced commercially, for example, could be as high resolution as that produced by military-intelligence surveillance satellites.

Political decisions clearly drive the field but economics becomes the other factor that helps drives what particular decision results from these political deliberations. This situation is especially true for those societies focused on fostering national economic development and competitiveness. For states such as China and India, especially in the beginning, those choices are hard ones given their large and growing populations and multiple economic deficits. Their budget choices become national zero-sum choices, a situation that normally translates into an uneven funding pattern especially during times of economic stress. Moneys are moved to cover the immediate crisis; natural disasters for example are profound shocks to the society. Even when economic good times follow, repairing past funding deficits can take precedence over pursuing new programs or expanding old ones. For both China and India, their respective national space programs are great sources of national pride but domestically such efforts must compete with other important economic priorities. Moving space technology to the next level may not rank higher than other social or other more immediate economic programs – immediate in terms of societal payoffs.

Technology, present and future

For many analysts studying space policy, whether globally or in the context of specific states such as the United States, Europe or China, technology exercises a particular fascination since technological capabilities are often defined as the *de facto* critical limiting factor upon a particular state's space aspirations. Those limits are in fact more fungible than previously considered. In the early days, that relationship was obviously true since no state until 1957 had demonstrated the capacity to successfully reach Earth orbit. The reality was that orbiting a satellite was technically possible earlier but political realities for years treated the feat as largely irrelevant. The critical task for both the Soviets and Americans was building accurate, reliable ICBMs. For the Chinese, the same military motivation drove their earliest effort; space activities were a clear second stage, a derivative of the first. So, for that reason, reaching orbit came as an afterthought for the national leaderships, although not for some leaders of the respective programs. Although for China, this second use – space operations – was obviously much more prominent from the start of their program, even then the primary priority focused on an effective missile program.

Wernher von Braun, for example, stated that the United States could have in fact reached orbit in 1956 but the Eisenhower administration was pursuing other goals.[33] Likewise, for the Soviets, Sputnik itself began the space race. In a Machiavellian sense, von Braun was correct in his judgment. The United States apparently pursued establishing the right of free passage of all satellites across the globe. International law prior to 1957 made national air space a protected region; intruders could be and were forced down or shot down by national military forces. Given this understanding, the question became how far up does national air space extend.

Once the first Soviet Sputnik flew in October 1957, the international precedent was set for free passage. In January 1958, the first US Explorer flew to orbit, further solidifying the precedent; both sides had now flown over the other's air space without challenge. The Soviets technically speaking continued to protest against satellite over-flights on principle but took no action to intercept. They could have tried and failed, preserving the general principle until their technologies grew capable of intercepting a satellite. Whether the Soviets considered this option is unknown, given that they were the first to establish the over-flight precedent. In time, the international precedent generalized and, as a result, the upper limit of national air space remains unspecified in law. This allows all states to fly their satellites in Earth orbit as long as there is no interference with the spacecraft of other states. From the US perspective, the importance was establishing the ability to fly spy satellites over Soviet territory without interference.

Anti-satellite weapons (ASAT) have long been technologically possible but both states in the end accepted free passage. China, lagging behind, benefited from this understanding. Ironically, this was contrary to their public view of the United States as an aggressor, especially with its continual over-flights of Chinese air space. China, in effect, benefited from the nuclear stalemate established by the two superpowers. Their early nuclear weapon efforts were not sufficient to be challenged either, although in the mid- to late 1960s the US government considered a preemptive strike against China's fledgling nuclear program, especially after the October 1964 nuclear test. That option was rejected for several reasons including difficulty.

That situation of relative political indifference to space launch changed dramatically in 1957–1958, not because the launch technology on either side, Soviet or American, suddenly improved but because of international politics. Nikita Khrushchev, the Soviet leader, employed Sputnik as political and military leverage against the United States. Achieving equivalent or better technological success became essential for the Americans in order to redress their perceived strategic inferiority. For states such as China, and actually for most states in the rest of the world, none of whom possessed the launch technology capable of reaching Earth orbit, this dispute was threatening but not critical. Even the Europeans did not achieve that goal routinely until the 1970s. Launch vehicles were prepared by several states but they were merely stunts or abortive efforts, rather than the result of sustained technology development.

Technology still remains the major impediment to achieving commercially and scientifically successful space operations, not because such launch technology cannot be successfully developed but because its failure rate remains significantly higher than its equivalent aviation component. Establishing commercial space operations demands reliability of both schedule and technology. The former refers to launching payloads on schedule in order to meet customer demands. Constant flight delays and rescheduling due to technical issues negated the US space shuttle's value as a commercial launch vehicle. The latter refers to losses or failures during flight, ranging from catastrophic destruction of the vehicle and its payload to delivering the payload to an incorrect orbit due to launch vehicle failures. These problems persist even among the most advanced space participants. The inaugural Delta IV heavy flight left the payload short of the desired orbit while the European Ariane 5 had several flight failures.[34] This is consistent with higher failure rates among new launch vehicles or new versions of older ones.

Space operations occur within a very unforgiving physical environment in which any errors of commission and omission are brutally punished by catastrophic flight or subsequent operational failure in orbit. How quickly new and more robust space technologies are developed depends upon national political choices based on available resources and technological capability. Government support, especially financial, and direction dominate the field generally. In the developed world, private funding options now exist but the core technologies employed usually first arose in the public sector. For China, that still remains true; the government clearly controls all the actions taken in the field. No true private sector exists within Chinese space activities although private businesses access and use those technologies. The government through its affiliates controls or influences all Chinese space-related activities. Corporations operating out of Hong Kong, for example, purchase comsats for launch possibly by the European company, Arianespace. This occurs due to US restrictions on technology transfer, which prevents US comsats being launched by Chinese lifters. The government often controls those companies or else it holds a major financial stake. The comsats can also be launched on Chinese Long March rockets although insurance concerns can force other vehicles to be used.

In some ways, the expanding global space commerce field significantly helps what are normally characterized as technologically weaker states by making space technologies more available. Back engineering represents one long-standing method employed by weaker or less technologically proficient states in order to advance. The formal international restrictions imposed earlier, such as in the Missile Technology Control Regime (MTCR), have weakened in recent years, although the United States continues to pursue restrictive policies regarding technology transfers.[35] Both China and India have been impacted by US efforts to control technology transfer, especially of space applications. Their subsequent behaviors were premised upon the principle that maintaining their national independence becomes critical for accomplishing their objectives. No other society is allowed a veto over these states' choices. Both China and India have been successful although a high price is paid for such independence. Advances come more slowly and with greater difficulty since basic technologies must often be in effect reinvented.

Space activities and their importance

Any consideration of China's space program must always take into account the world situation confronting China – while also incorporating the dramatic upheavals that have characterized Chinese domestic politics. Space activities are normally considered dual-use in nature. This simply means that essentially the same space technologies can be readily employed for both civilian and military uses. For example, the same rocket technology can lift either a civil-commercial payload to orbit or deliver a warhead on target. Unfortunately, but realistically, it is the latter use, the military, that most often first motivates governments to develop such technologies. Civilian uses flow from that technology but do not normally engender the same degree of government support, especially financial. The Europeans in the form of the European Space Agency (ESA) are often considered an exception, as were the Japanese with their constitutional limitation on military activities. Those exceptions are evaporating under the pressure of post Cold War security events.

In actuality, space technologies could equally well be characterized as being triple-use in nature rather than simply dual. The third facet is prestige, which arises from the employment of space technologies to enhance the state's and governing elite's international and domestic prestige. This latter characteristic is particularly important for regimes or governments insecure in their possession of power or threatened by potential external enemies. The early Soviet–American space race exhibited all of those characteristics. Both states perceived themselves as threatened by the other while both national leaderships came to see successful space activities as essential for their remaining in power.

For example, Nikita Khrushchev found "space firsts" extremely useful in keeping the United States off-balance internationally while at the same minimizing possible perceptions of Soviet strategic inferiority. Employing various forums including the United Nations, Khrushchev trumpeted Soviet leadership in space activities as truly symbolic of its coming international dominance economically and politically. In his analysis, the United States was cast as the Soviets' inferior militarily and economically and, according to Khrushchev, it was a situation that could only inevitably and progressively worsen.

American leaders were initially dismayed by their apparent public impotence in challenging Soviet space leadership and embarked on an expensive space race to catch up. President Dwight Eisenhower was not convinced of that reality – his view remained that the United States was militarily secure. Congressional members within his own political party were less sanguine and forced Eisenhower to respond. The apparent Soviet lead in conducting space activities however did not signal their dominance with regard to nuclear weapons – a fact brought graphically home to the Soviet leadership during the 1962 Cuban Missile Crisis. American strategic dominance was not in fact in jeopardy during the early space age but space activities became the most public manifestation of the superpower rivalry, a battle the Americans were very publicly losing. Peaceful space activities' importance was as a surrogate for direct military confrontation – a competition peaceful in nature unlike the clash of nuclear-tipped missile forces. Aside from the two superpowers at the time, no other state could engage in the competition. The space race occurred at a level of intense economic effort unimaginable for any state other than those two. In fact, by the late 1960s, both states found the economic exertion unjustified while both downscaled their civil activities. However, expenditures for military space continued and grew, illustrating its fundamental importance for the superpowers.

For China, space activities also possess great importance as a symbol of its advancing technological, economic and military capabilities. In an interesting sense, space activities still substitute for military activity by demonstrating convincingly that China has now clearly entered the realm of a major military power. At one level, the space symbol now outruns the realities of Chinese military capabilities, which remain mostly regional in their scope and effectiveness. China's nuclear arsenal grows but remains qualitatively behind those of Russia and the United States.

For China, however, the importance of space activities is accentuated because of the extremely poor economic position from which it first initiated the pursuit of space activities in the 1950s. China's launch of humans into outer space on October 15, 2003 now places it at a level unmatched by any other state, even ahead of the Japanese and the Europeans who opened their space efforts from national positions of much greater economic and technological resources. Both had to rebuild from the devastation of World War II but they needed to replace lost materials while the human expertise and experience base was still available. For China, the struggle is to build both the material base and also an expertise base along with experience using modern technologies.

Rebuilding China

China's space program did not arise out of a vacuum but rather was initiated within a specific society, one that has endured great trauma and travail for more than a hundred years. The first three-quarters of the twentieth century was a period of great turmoil within China. The last Qing (Manchu) imperial emperor was dethroned in 1911, although China's power had been systematically debilitated for more than four decades due to the repeated incursions of the European colonial powers and a failing dynasty. In 1911, a republic led by Sun Yat-sen replaced the monarchy, but Sun Yat-sen himself was replaced by General Yuan Shikai who reestablished the monarchy in 1915 by proclaiming himself as the emperor. After Yuan and his regime died in 1916, China entered a period of civil war during which different alliances of warlords struggled for dominance. This warlord politics did not end until 1928, when General Chiang Kai-shek preliminarily put China under the control of Nationalist-led government. But Chiang's regime proved extremely unstable with the eruption of new civil wars with regional warlords and the emerging Chinese Communist Party (CCP) led by Mao Zedong. The incessant wars disrupted economic growth especially in urban areas.

Chiang attempted to implement some reforms to strengthen China, but corruption within the party along with continual strife with the CCP and the warlords sapped its resources. By the early 1930s, the Nationalists appeared to be dominating the CCP. For example, in October 1934, the Red Army led by Mao Zedong embarked on their long retreat to safer, less accessible regions within China, known as the Long March. This effort did not end until October 1935. From the beginning of the 1930s, Japan began making its territorial demands on China more openly and aggressively. China had already been attacked by Japan in September 1931, when they seized control over Manchuria. In 1937, the second Sino-Japanese War broke out and the Nationalist government had to fight against the Japanese. Although Chiang Kai-shek was forced to cease hostilities against the CCP and to form the second Kuomintang-CCP United Anti-Japanese Front in December 1936, the distrust between the two never ceased and conflicts escalated after 1941 and consequently Chinese society was severely devastated.

Other states, led by the United States, ultimately defeated Japan, allowing the civil war with the Communists to intensify. In 1949, the Chinese Communists seized power on the mainland under the leadership of Mao Zedong whose rule until 1976 was characterized by a series of political movements which were aimed at consolidating his personal control of the party and ultimately the state. The Nationalists led by Chiang Kai-shek fled to Formosa (or Taiwan) where they consolidated their power. The Nationalists saw Taiwan as their new seat of power while they plotted their return to the mainland, the start of the dispute over the two Chinas.

The People's Republic of China (PRC), established on October 1, 1949, embarked on consolidation of their power on the mainland. This process was interrupted by China's involvement in the Korean War. North Korea, with the consent if not encouragement of Joseph Stalin, struck across their common border with South Korea in June 1950.[36] This action brought the United States militarily back into East Asian affairs when it led a

United Nations (UN) coalition against North Korea. By December 1950, North Korea was in full retreat after the Incohin landing as UN forces closed in on the Yalu River, the border with China. Rather than formally declare war on the United States and UN, Chinese armed forces entered the conflict with the result that by 1953, the conflict ended in stalemate. The Korean War proved to be a catalyst for heightened Chinese–US hostilities across the spectrum of activities. The United States extended its military protection over Nationalist China and embarked on a series of mutual defense treaties with Japan and other Asian states. One multilateral example was the 1954 South East Asia Treaty Organization (SEATO), which was established just as the French retreated from Indo-China. This treaty plus the alliances with Nationalist China and South Korea fed Chinese concerns about being surrounded by enemies. That perspective was further justified by the earlier example of the Soviet Union surrounded by bourgeois enemies.

From China's perspective, the United States with its nuclear arsenal and hostile posture presented a very real military threat. What was less clear to China's leaders was the US willingness to militarily intervene on the Asian mainland after its experience in Korea. US attention was focused upon Europe and their fear of a Soviet invasion. The North Atlantic Treaty Organization (NATO) alliance dominated US attention and commitments. That shift in attention meant that once the Korean War stalemated in a truce, China had the opportunity to focus on its internal development including economic reforms such as redistribution of land and attacks upon the landowners. In 1953, China embarked upon its first Five Year Plan (1953–1957). The goal was to stimulate industrialization and complete land collectivization. During this period, there arose the Hundred Flowers Campaign – an aborted attempt to liberalize the political climate. The phrase, "Hundred Flowers," came from a slogan: "Let a hundred flowers bloom, let the hundred schools of thought contend." This aim was claimed by the CCP, but many people believe that Mao's real aim was "to lure snakes out of their holes."[37] People, particularly the educated, spoke out their true thoughts, but the government promptly repressed criticism. After a period of political turmoil, China continued its push for industrialization.

In 1958 Chairman Mao along with his political supporters concluded that China's pace of economic development must be accelerated. The result was a new endeavor best known as the "Great Leap Forward." The goal was to ideologically motivate the workers and to organize them into "people's communes." Each commune was to constitute a self-sustaining community. The symbol of this effort internationally was the backyard pig-iron smelters. The expected efficiencies did not occur, with the result that Mao left his position as Chairman of the PRC while retaining his position as Chairman of the CCP. At the same time, ideological and personality conflicts with the Soviet leadership led by Nikita Khrushchev escalated to a crisis, ending with the Soviet Union terminating its program assisting China in nuclear and missile development. China was effectively isolated, meaning that independent development of those technologies became a priority.

In October 1964, China held its first nuclear bomb test – a major step but one that aroused American anxieties concerning China's military potential. By 1964, the Vietnam War was being escalated by the United States, posing a potential threat for China which was considered to be a sanctuary for North Vietnam. During this time period, the US government seriously considered the possibility of preemptive strikes against China's nuclear facilities. That option was rejected in the end but relations remained strained. China was in the politically awkward position of becoming militarily stronger (acquiring nuclear weapons) without the means of delivering the weapons against potential enemies, the United States and Soviet Union. Even if missiles were developed, a Chinese version of the earlier US "missile gap" would have heightened their vulnerability. The Soviets and Americans possessed large missile forces (in excess of 1,000) plus long-range bombers. All of this added to the urgency of the Chinese missile program. However, domestic political upheavals intervened to delay technological development from accelerating.

Mao Zedong by the early 1960s was in partial political eclipse. After the collapse of the Great Leap Forward, Mao found that he could not totally control the bureaucracy and the party. He complained that he could only get a slight majority at the conference decisions in early 1966.[38] To seize his power back, Mao launched the Cultural Revolution and purged his potential political rival Liu Shaoqi and millions of Liu's followers with the denouncement of "bourgeois-liners," with the backing of the PLA led by Lin Biao. What followed was a decade-long period of upheaval labeled the Great Proletarian Cultural Revolution. For the first several years, a battle ensued between the young people mobilized as the Red Guards to directly and physically challenge the existing order including many elements within the CCP. Key party and government leaders including Liu Shaoqi, chairman of the PRC, and Deng Xiaoping, Secretary General of the Central Committee of the CCP, were purged from public life, forced to engage in self-criticism and redemptive manual labor.

By the spring of 1969, the greatest disturbances had ended with the Maoists seizing power. Mao Zedong was again designated supreme leader while Lin Biao, the Defense Minister, assumed the number two position. However, the PLA (one key power center) was split over the ideological push and China's increasing isolation, given that military clashes occurred along the border with the Soviet Union. In September 1971, Lin Biao attempted a political coup, which failed, and Lin died while reportedly fleeing China for political refuge in the Soviet Union. Subsequently, President Richard Nixon visited China in February 1972, creating a balance against the Soviet Union's expansive rhetoric in the Brezhnev Doctrine which stated that the Soviet Union could intervene in socialist states' domestic politics to redress deviations from Soviet leadership. "When forces that are hostile to *socialism* try to turn the development of some socialist country towards *capitalism*, it becomes not only a problem of the country concerned, but a common problem and concern of all socialist countries."[39] That doctrine was a direct response to unrest in Eastern Europe as symbolized by the invasion of Czechoslovakia in August 1968, ending the "Prague Spring." Lin Biao's attempted coup heightened the dangers inherent in domestic strife.

The Cultural Revolution was gradually rolled back, as symbolized by the rehabilitation of Deng Xiaoping as vice premier in 1973. Premier Zhou Enlai led the more moderate elements but that effort was cut short by his death in January 1976. Mao Zedong died later that same year in September although turmoil characterized Chinese politics until the purge of the vestiges of the "Gang of Four" and other more radical elements. Deng Xiaoping in 1976 was temporarily driven from office again; his return in 1977 signaled the demise of the Cultural Revolution in a real sense – a fact ratified by the Eleventh National Party Congress in August 1977. A final showdown in 1978 saw the dominance of the moderate wing of the CCP, PLA and government generally.

In December 1978, China set forth on the process of modernization and development known as the "Four Modernizations," announced previously by Premier Zhou Enlai in 1975. The four included the modernization of industry, agriculture, science and technology, and national defense. Economic progress was now more important than the Maoist goals of class struggle and permanent revolution. Deng Xiaoping consolidated his power through his position as chairman of the CCP's Central Military Commission, a position that kept him in touch with and in some control over the PLA. Also, China during this period established formal diplomatic ties with the United States, fought a brief war with Vietnam over shared borders, and ended its thirty-year treaty of Friendship with the Soviet Union.

Deng's policies included major changes (even reversals) of Chinese policies. The goal was fundamental change especially economically in an effort to make China truly a player on the world stage. Developing technologies and expertise in those technologies demanded great effort while opening up China's economy to international capitalism as a source of investment in China. What developed in China was a hybrid state, partially state-run and the other part capitalist, which had been banned earlier. The result was a situation of intense though unevenly distributed economic growth, one consequence was that China began its recovery from the partial chaos of the earlier years.

Securitywise, China also began its growth as a military and economic power – a position that holds interesting consequences. Formal relations with the United States contrary to Chinese wishes did not signal withdrawal of US protection for Taiwanese independence from the PRC. China's economic growth included heavy use of exports including its Silkworm missiles to Iran and Dong Feng-3 missiles to Saudi Arabia. China in time agreed to cease such sales. The collapse of the Soviet Union in 1991 had ended any immediate military threat from that source. For China, the United States remained the only immediate threat and only in the event of a war over Taiwan. The 1989 Tiananmen Square democracy demonstrations and their crushing by the government inflamed animosities with the United States. Human rights questions remain a constant difficulty, especially with regard to the US Congress, which crimps a president's discretion. In the late 1990s, a controversy arose over the possibility that US companies provided technical information to improve China's launch program. The controversy was mostly for domestic US political consumption but one consequence was to limit the ability of US comsat operators to fly payloads on Chinese launchers.

Thus, by 2005, the United States and China were officially open to each other to a degree unprecedented in recent history. However, each perceived the other as a likely Asia-Pacific rival militarily and economically. Cooperation continues on issues such as North Korea and its nuclear ambitions but both countries' militaries saw each other as positioning themselves for possible conflict. China perceived the US national ballistic missile defense program as positioned to nullify the Chinese nuclear missile arsenal. From their perspective, the United States was rendering itself immune to retaliation if it attacked China. To the Americans, the Chinese were threatening with their talk of taking down US military satellites – the backbone of its global reach militarily but also a very vulnerable segment of its forces. In response to perceived American imperialist tendencies, the Chinese are reasserting their ties with the Russian Federation.

For China, its space program has grown from a symbolically important act to part of the foundation of Chinese economic, military and political power. As will be discussed, China has moved far beyond its earlier status to one of world leader. The Chinese space program has been part and parcel of that change, benefiting from the improvement but also adding technological and economic luster to the entire enterprise. Space is no longer the toy of the rich but a fundamental part of China's developmental plans.

Conclusion

In Chapter 2, we provide an overview of the general development of the space realm from the earliest program to later entrants. Chinese efforts are part of the historical flow of events. By its presence, China impacts the entire field's development and future directions, changing what is meant by the concept of a national space program from a luxury to a necessity. The path as it is traced out is not easy but the rewards are seen as great enough to support the desire.

Chapter 2

China as space follower and leader

Introduction

In this chapter, China's quest for significant achievement in space activities is placed in the larger international space context against which its program can be evaluated. This approach to studying the Chinese space program is different from that attempted earlier by other scholars because their analyses fail to embed the program in the global technological context within which all national space programs operate. One must complete such an analysis in order to more objectively evaluate the Chinese efforts rather than making their efforts appear excessively unique or, conversely, erroneously underplaying the strong domestic roots upon which all national space programs are firmly grounded.

The study of space policy has too long been plagued by excessive fascination with the uniqueness argument; each state is presumed to reinvent the "space wheel" as it were when it embarks on a national space program. After fifty years, there exists ample historical precedent against which all newer space programs can be evaluated. This is not to underestimate the magnitude of the technological and financial task to be accomplished by each state. Each in its turn must marshal the vast resources (financial and technological) and expertise necessary to design, build and operate truly independent space technologies.

Access to space technologies can now in many cases be purchased (at least for in-orbit use) by new participants. China, until the 1980s and especially the 1970s, lacked that option which is becoming presently available. Therefore, technology independence, a goal already strongly desired by the Chinese elites, was forced upon China. India has encountered the same issue albeit less intensely than China, although to the Indian government the intensity was sufficient to delay acquisition of critical technologies. US policy was often relatively indifferent to India but not regarding China, which was perceived as a hostile state, a perception in fact held by both parties.

In response, China has assertively employed its status as the great-untapped market to gain concessions from commercial space operators and their government sponsors. Despite the relative poverty of much of China, the markets available in terms of total population make competition to access them vigorous. China has proceeded through a combination of partially government-owned companies and international corporations to open itself up to those space applications supportive of its national agenda. The Four Modernizations alluded to briefly in Chapter 1 remain very relevant to the development of the Chinese space program. Prestige is nice but you cannot eat it, meaning a strong thread of practicality runs through Chinese endeavors. That sense of constraint was not exhibited by the first two space pioneers, the United States and the Soviet Union, but their efforts lagged over time.

National pursuit of space activities

Pursuit of space activities by different states has to this point in time always been the product of profoundly political judgments rather than purely technical. This political foundation can be found both in terms of the states' original decision to initiate that effort and, subsequently, the degree of assertiveness with which that national goal of independent space activities is pushed forward, often in the face of immense obstacles. Some obstacles are mostly physical in nature but more often the overcoming of the obstacles represents the politics of choice as played out in a specific national political context. That is what makes Chinese choices regarding space activities different from those of other states.

Initial differences in available economic and technological resources and changes in political leadership can significantly alter what is possible. However, that surface national uniqueness should not obscure the regularities that underlay the historical processes embodied in the development of the space age. All states that pursue acquiring assured access to space must proceed through certain stages including first acquiring relevant technologies. How swiftly each state accomplishes these essential tasks depends upon the choices made by its leadership.

These political choices must be reaffirmed constantly because the goal is often distant, while newer demands press for resources, draining them from existing programs. In the Chinese case, as will be described, the decisions to push forward had to be repeatedly reaffirmed. Rarely is the choice one of cancellation or continuation but rather how fast the program should be pushed. For China, given its resource deficits initially, its decisions to continue or expand the program were always difficult.

Because all other space states track behind the two original space participants, they begin with one certainty in mind: accessing and using space technologies is not only possible but has been done. One speaks of "standing on the shoulders of giants" in pushing forward the frontiers of science – the same phenomenon occurs with regard to technologies. Today, a state knows it can be done and, more critically, what technologies and resources at a minimum are required.

What blinds many analysts to this essential political reality is their general fixation on the technologies involved. Technology's costs and complexities, along with the very real difficulties in making space technologies work successfully, repeatedly divert attention from the essential political dynamics underlying the entire space development process. Politics are considered by space practitioners and enthusiasts as arbitrary and ignorant outside interference rather than what they are in fact: enablers of space activities. In truth, often both facets occur but part of that interference reflects the unruliness of politics in any society, whether democratic or more authoritarian in nature.

Authoritarian governments often appear more monolithic than they are in fact. But, their politics are often a constantly changing combination of personalities and factional politics, as has occurred in China since the establishment of the People's Republic of China in 1949. In effect, court politics determine what is done or even possible. Elections in effect determine within the elite who holds power and for how long. Otherwise, repression or civil conflict breaks out, often with enormous damage to the state and society. Within China, the opaqueness of its elite politics often puzzled outside observers by exaggerating the degree of agreement. The Cultural Revolution from 1966 until 1976 roiled Chinese politics and society, reversing that view for a time to one of chaos. When order was restored, political stability became the regime's mantra even as dramatic changes occurred throughout China's economy and society.

Democracies are visibly disruptive only because most decisions occur in a very public political process, with some degree of compromise the usual outcome. Strong tides can run in democracies, only to reverse themselves equally quickly. As a consequence, the attention spans of government officials are often very short-term in nature; so long-running programs such as space activities suffer as the tide ebbs. The US, European and Japanese space programs have all been adversely and positively impacted by such shifts in the currents of politics within their societies.

Engineers and technologists too often become fixated on the "sweetness" of the technical solutions developed, a perspective that rubs off on space analysts whose view should be wider in scope. Space technology development is analogized as the Darwinian struggle of humans against nature operating at the extremes of technology possibilities, "cutting edge" in other words. What becomes lost is any real sense of the disproportionately large societal investment constantly demanded from often distracted and cross-pressured political authorities. Space activities are not cheap and, for a new participant, the costs are constantly escalating with few immediate results. This cost curve impacts all space participants, private or public, economically developed or underdeveloped.

Within this larger perspective, China must be judged no different from any other space-aspiring society. The Chinese program's ability to push forward despite repeated technological failures and episodic resource shortages reflects ultimately the Chinese leadership's political determination to succeed. As a further example, Brazil continues to pursue space launch despite three major mishaps including an explosion on August 22, 2003 which killed twenty-one personnel.[1] Technical personnel doing the work are often plagued by interference from the political authorities but that is the price that is paid since those efforts are dependent on public funds. Thus far, in those states where a robust private space sector exists, that private sector has often been very active but under generally tight guidance of the program's officials. There has not existed a purely private space program yet, one moving from developing independent launch to spacecraft. Private launch companies do exist but all employ formerly government launchers or demand public funding for upgrades. Comsat companies may be marginally more independent but most got started using government contracts.

When the possibility of reaching Earth orbit and beyond became real, the various states interested in space activities did not initiate their activities on an equal footing. Some states, especially the developed ones, started their programs clearly more advantaged in terms of resources and technological capabilities than others. Political determination in and of itself is insufficient to immediately overcome all major technological and financial deficits. The emphasis is upon the word "immediately." For most states outside the charmed circle of the two superpowers and their closest allies, those capabilities largely did not exist in the beginning but rather their programs were deliberately created as the result of government choices.

Generally, dedicated political support has proven absolutely essential for achieving whatever space activity was possible for most states. The process of establishing an independent national presence in space activities remains a long-term one rather than a quick fix, as several states have discovered to their dismay. Simply scaling up already existing military rocket technologies, for example, did not translate into immediate launch success. What is often forgotten is the large number of launch failures that occurred in the United States, failures reported at the time, and the Soviet Union's more secretive failures.[2] The United Kingdom for example was involved in the Europa project in the late 1960s and early 1970s. That joint project represented Western Europe's first attempt to establish a space launch capability truly independent of the United States. France had earlier, in November 1965, launched Diamant A to orbit, but subsequently joined the European Launcher Development Organization (ELDO) which pursued development of the Europa rocket. The Europa rocket was the first example of an "off the shelf" spacecraft, a fiasco in the end. Europa was built from several components including the British Blue Streak first stage, followed by the French Coralie second stage, and a third stage to be built by Germany.

The immediate project failed miserably, not because the six primary participant states involved, plus Australia as an associate member (all economically developed), could not work out the technology but because the political organization of the effort and therefore the technical arrangements ultimately became too cumbersome and awkward to work in practice.[3]

However, once the organizational tools were put in place – first, the European Space Agency (ESA) and eventually Arianespace – European efforts became more focused and systematic, even though the entire development process remained marinated in politics. The organizational structure of ESA was set up to accommodate the interests of all participants, allowing all participating states a share in the action but stratifying that participation. Being a confederation in structure, resolution of issues takes time and is often plagued by slow starts, as all the critical participants must be brought on board, especially the core states of France, Germany, Italy, and the United Kingdom more recently. Once decisions were agreed to by those involved, implementation appeared to flow almost automatically to program success.

In fact, for a short time, there was some speculation that the European Space Agency (ESA) was somehow immune to some political problems endemic to the US space program with its false starts and continual budgetary issues.[4] The Soviet-Russian program was also considered immune to financial issues but that perception vanished with the collapse of the Soviet Union. Once decisions were made, they were carried out within ESA because their system was built on multiyear and programmatic budgeting rather than annual budgeting. The latter situation facilitates continual reconsideration of earlier budget decisions – nothing is ever considered truly final. That optimistic perspective regarding ESA had evaporated by the late 1990s when European economic uncertainties forced slowdowns and restructuring of programs in order to conform to ever tightening budget restrictions.

In fact, there is no magic wand that leaves a national space program exempt from the normal rules of domestic politics including budget constraints imposed by constantly changing national economies. European ministers were forced to make budget tradeoffs; space activities did not necessarily do well in those decisions. No major program was immediately shut down but aspirations were often tempered. The original American experience with the Apollo program beginning in May 1961 influenced perceptions of what could be done. For a brief shining moment, the pursuit of certain space activities soared beyond the boundaries of normal budgetary politics. What was generally forgotten was how quickly the Apollo project fell back to earth amid the realities of domestic US budgetary politics. The Soviets likewise appeared immune initially, although by the end of the 1960s even their space program entered a less frantic stage of space activity. The political will was gone for mounting the frantic efforts of the early 1960s.

Why a state pursues space activities and at what level remains an individualistic decision – one that individual states approach through the lens of their particular national perspective. That individuality does not mean that a more generalized perspective cannot be created for analyzing the field. As indicated in Chapter 1, the specific factors affecting national space activities are a combination of politics, economics and technology, including international political and military considerations. These latter considerations, especially the military, were especially pressing during the space age's earliest days but neither can be separated out as the single causal factor.

Also, specific national motivations shift over time in terms of their intensity. For example, commercial considerations might be the primary driver at first but subsequently military threat perceptions might become more critical. One can see that aspect playing out with regard to the European Space Agency, which is evolving toward applications whose uses are more ambiguous but now include possible military applications. Other states, including China, began their programs originally with a military purpose, given the perceived security threat posed by the United States and the Soviet Union. Over time, other uses grew more critical or central to what the space program pursued. That does not mean that the military facet disappeared or became insignificant; rather it was incorporated into the normal expectations held by the state. Conversely, Europe and Japan began their space programs emphasizing scientific and commercial applications but events and their international situations are raising the military facet to greater prominence.

Constantly overhanging all national space programs is the military dimension, since national security remains a particularly potent justification for the large expenditures demanded. However, one should be aware that military space activities build logically out of peaceful space activities, so that pursuing military space activities first and then shifting to peaceful, or doing it vice versa, constitutes merely a change in emphasis, but not technologically speaking. The military dimension, if an initial justification, adds to the intensity of the effort.

Rationales for initiating and expanding national space programs can be considered to fall into distinct clusters, as depicted in Table 2.1. It is argued above that specific early political events lay the basis from whence flowed the specific national programs and activities operating today for a diversity of space participant states. For example, the Soviet launch of their first satellite into Earth orbit on October 4, 1957 galvanized US space activities. The military threat posed by the Soviet satellite and its launcher (a surrogate for ballistic missile capabilities) generated an intense political-military response by the United States. The specific time frame within which a particular state initiated its pursuit of space activities had a direct impact upon how their space program was formulated and pursued forward in time. The earliest space programs had the strongest military roots. China

Table 2.1 Rationales for national space policy initiatives

Activity	Beginnings	Present	Future
Military	Soviet Union/Russia, US, China	Force enhancement Force support	Space control Force application
Scientific	Soviet Union/Russia, US, China, Europe, Japan, India, Brazil, Israel	Space science, Earth science, Environment	Search for life
Commercial	Russia, Europe, Japan, India, Brazil, Israel	Commercial applications	Economic competitiveness
Human spaceflight	Soviet Union/Russia, US, China	Independent & cooperative human spaceflight activities	Human spaceflight activities
	Canada, Europe, Japan	Cooperative human spaceflight activities	Independent human spaceflight activities

chronologically falls into that group even though its early efforts were often out of the vision of the two major space powers. China's potential, however, worried both powers who recognized that China's motivations were similar to their own.

The result is that space activities can be conceptualized in terms of several relatively distinct waves of program initiation since 1957. The Soviet Union and the United States were the original first-wave national space participants whose motivations were dominated by military-political considerations, especially military, but crossing the entire spectrum. Both states held their military space efforts in closest confidence, with misdirection and secrecy the rule.

The peaceful facets of their programs, scientific and human spaceflight, became the public face, but those were often considered a diversion from the most important sector, the military. For example, President Eisenhower in the 1950s saw civil space activities as a diversion from the real task of developing accurate and reliable intercontinental ballistic missiles (ICBMs). As a consequence, their space programs became all-encompassing in that each state felt it necessary to compete across all realms of space activities, except commercial for the Soviets, although commercially equivalent technologies were developed for domestic Soviet use. Communications satellites (comsats) were the earliest obvious commercial application – one that was also extremely useful for military purposes.

For those states, the military justification provided the requisite funding and political support for their entire space effort. Whenever that military justification was lost or became irrelevant, political support and consequently the pattern of strong funding declined precipitously for their civil space efforts.

The historically best-known example of that reverse phenomenon was the rapid dismantling of the US Apollo program once the political-prestige objective of landing humans on the lunar surface and returning those individuals successfully to Earth was accomplished. The Apollo program was a particularly clear example of a state space effort driven by short-term political and national prestige considerations. The program effectively evaporated once political support was withdrawn. In fact, if President John Kennedy had been alive to serve a second term (he was assassinated in November 1963), there is some evidence that the Apollo program might have been cancelled before success was achieved, or changed into a joint Soviet–American effort.[5] Rather than being a bold journey into the unknown future, Apollo became a mundane political gesture whose life ironically was extended by its progenitor's death.

China's leadership at the time of the Soviet feat of first launching satellites into Earth orbit (October 1957) was deeply impressed by the rocket technology with its obvious national security and prestige benefits but at that time clearly lacked the technological base to immediately follow suit. Their own efforts at establishing a military missile program were still in their infancy. That reality meant that the national leadership did not at first initiate a distinctly Chinese space program since they were largely dependent upon the Soviets for technical assistance. Whether the Soviets would have been very supportive of a truly significant Chinese space effort is unknown given the two space superpowers' penchant for retaining control over militarily important technologies. The entire question of assistance became moot when an intense political schism arose in the period 1959–1961 between Mao Zedong and Nikita Khrushchev; the differences incorporated political, strategic and personal dimensions given the centralized nature of both regimes.

So, the reality of the Chinese space program was that their effort grew out of that same sense of military necessity along with the clear ancillary value of fostering China's international political prestige in a manner associated with the first two space participants' exploits. China's twin motivations clearly mirrored those of the United States and Soviet Union. This military security/international prestige emphasis is what leads to China being classified here with the first two space participants whose programs foreshadowed theirs significantly in time. This relationship is often ignored, with China's effort often defined as representing something totally new rather than as part of an established historical developmental process.

Table 2.1 indicates that both the United States and the Soviet Union, now the Russian Federation, pushed their military space programs aggressively in order to improve their military's effectiveness. Force enhancement involves the provision of communications and critical reconnaissance and observation data in a timely manner to military forces in the field. At the earliest stages, that assistance could be as simple as global weather data since weather can have devastating effects upon military operations. Likewise, acquiring targeting information and timely knowledge concerning enemy movements, locations and possible attacks were critical items for which large sums were spent willingly. Early warning satellites became critical components in those efforts to enhance the survivability of their nuclear forces. The enhancement function has grown even more critical as modern militaries move toward a web-centric approach to warfare.

At first, China's primary difficulty lay in the fact that the state lacked the requisite resources to pursue those activities at anywhere near the same level of intensity as the first two space powers. Combined with China's serious technology deficits, that resource factor led to underplaying the public visibility of China's military justification for their space program internationally but not with regard to regional military competitors. India and Japan were well aware of China's military goals but for various reasons – available resources in India's case and national politics in Japan – their programs lagged behind China. Japan was more visibly engaged in space activities, especially through cooperative international projects, but their military significance was long minimized by the Japanese. For India, resource issues plus its technological isolation due to the Missile Technology Control Regime (MTCR) restrictions impacted its ability to respond quickly.

For the second-wave space participants (1960s and early 1970s), factors of resource constraints and possible technological deficits dominated their internal debates; the primary public rationale expressed became scientific and, in a more long-term sense, commercial in nature. One major explicit political goal of pursuing space technologies was to build up their domestic infrastructure technologically and economically. Politically speaking, the states in the second wave tended to be technologically advanced in many areas other than space activities. Examples include Japan and the major European states. Their lateness in space reflected their earlier policy choices which were now back in play. Basically, those governments were unwilling to accept permanent inferior or subordinate status relative to the first two space powers. This disgruntlement was obviously more visible among US allies, where dissent was possible, compared to the Soviet bloc. ESA became the most public manifestation of that European desire for greater independence from the United States. This aspiration was also shared by Canada which joined ESA as an associate member, but was constrained by geographic distance and resource limitations.[6]

In the 1970s and 1980s, China was not perceived as the second wave's equivalent in terms of technological development. Given its major technology and resource constraints, however, more conscious political effort was expended by China to prioritize its space activities and to make hard policy choices regarding what space activities should be pursued once orbital access was achieved in April 1970. In a critical sense, these second-wave states are clearly different from the two original space participants, whose programs responded in an almost Pavlovian fashion to the other. Whatever the other state pursued with regard to space activities, the other felt obligated to duplicate or to move beyond to the next level. That attitude essentially explained the early Soviet and American constant obsession with space "firsts." The question for those two states was not what could be afforded but what was deemed politically necessary. At least for a time, politics trumped any deep rationality about cost or orderly progression of missions. With both states, this obsession did not go without internal criticism but the political imperatives of military security and international prestige were considered more pressing. President Dwight Eisenhower established and expanded the US program only under great political duress since he personally did not perceive the civil space area to be of sufficient importance militarily or otherwise to merit the extraordinary resources being demanded.[7]

Conversely, for those states in the second wave, costs were considered very real because no overriding political justification existed by which to beat back domestic opposition objections that the resources should be used elsewhere. Tight national budgets and pressing social and economic needs drained the political intensity from their civil space efforts. None of them were going to operate a national space program equivalent to that of the superpowers.

Ironically, both the United States and the Soviet Union subsequently reached the point where economics did matter but their thresholds for such second thoughts were much higher than those of the second-wave states. Technologies or the capabilities to build such space applications clearly existed in several industrial states outside the original two. What was lacking in those states was the political intensity that a perceived military threat brings to the debate. Both Europe and Japan were protected by the US nuclear shield.

China from its initial tentative steps had the political urgency but not the resources to move forward quickly. That deficiency meant early Chinese efforts included several false starts before steady progress became the norm. Many early starts were aborted due to domestic political disputes over national priorities, a common occurrence for government space programs. The reason is simple: space programs take time and are often overtaken by more recent events either internationally or domestically; or, the national economy expands or contracts, changing the revenue stream available for such exotic ventures.

As a result, second-wave states set off in pursuit of national space activities where their efforts were in principle more moderate financially but also more subject to expanding domestic political demands. One manifestation of this vulnerability was their active interest in international projects, one device by which their consciously limited technological and financial resources could be stretched further. Such cooperative activities are only truly possible in nonmilitary space areas, so that their early focus was on space science and commercial applications. The Europeans' collective space effort through ESA has been the most completely documented but the overall pattern has been fairly clear.

Japan followed Europe, albeit at a distance in terms of their political intensity, but it initiated and sustained a technologically independent space program through the medium of university-based space science. Those efforts included developing launch vehicles to carry national payloads to orbit and beyond. Japan, like Germany, also had to overcome the legacy of World War II. Japan's defeat led to its forced disarmament and the imposition of a constitution that stated that Japan would not engage in military activities or activities supportive of military activities. Japanese space technologies suffered as a result since all such technologies clearly could be used in a military fashion even if no weapons were involved. Once Japan developed a peaceful rationale for its space activities, development picked up momentum especially its launch vehicles, the latest being the H-IIA. Further changes came to Japan's space program as their concerns about North Korea with its missile program and possible nuclear weapons grew. North Korea's physical proximity accentuated Japanese insecurities especially after their 1998 missile space launch over Japan.[8]

That launch accentuated Japan's vulnerability in terms of early warning and possible defensive action. China's growing space program has added urgency to the Japanese efforts. Their reactions however feed back into the Chinese space effort – a competition is growing between the two. The impact is greatest on the Japanese space program, while China remains concerned with maintaining its current advantage *vis-à-vis* Japan. In a sense, this tracks the early space age when the two protagonists were more evenly matched and not disabled by political constraints similar to those impacting Japan.

As has been indicated earlier, one should not be deceived by the "peaceful" focus of these second-wave space activities. Obviously, the concept of dual-use technology applies directly to most space applications. The reality is that the same technology can be employed for civil and military purposes. The earliest example remains the rocket, which carries either a payload to orbit or a warhead to a target across the ocean. For this reason, the earliest experiences in space launch usually became extensions of ongoing work on missiles, i.e. launchers for the purpose of delivering warheads to a target. Likewise, satellites take pictures from orbit; those pictures can be of weather patterns, for environmental monitoring, or of military installations and troop movements. Nothing in principle distinguishes the two applications except the user's intent. That intent can obviously change quickly. Japan is an excellent example of this potential. So, conversely, space programs pursuing the technologies necessary to conduct space science or commercial operations provide a backdoor for all states directly into the realm of useful military space activities.

That reality is why both the Soviets and the Americans from the beginning attempted to control their major allies' access to Earth orbit; both failed although for different reasons. One should note that state participation in space activities does not axiomatically lead to a military space program, but the step cannot be considered far off or as difficult to accomplish as many assume. National leaderships must deliberately choose their future with regard to the military space question. In that sense, these states perceive themselves as freer than the first-wave states for which military issues were so overwhelming and still impact their space programs.

With the Cold War's decline and the eventual disappearance of the Soviet Union, second-wave states are now finding military space applications particularly enticing, in part because of their changing relationship with the United States. Formerly, these states were effectively frozen in place or at least severely limited in their flexibility regarding their foreign policy choices. The United States as a self-proclaimed but generally singular superpower finds its erstwhile allies more reluctant to follow its lead. On the other hand, their comparative military weakness especially in high-tech warfare is forcing their space programs to become more focused on military applications. The two Gulf Wars (1991 and 2003), along with the military actions in the Balkans, especially the Kosovo air campaign, repeatedly demonstrated the wide gap in capabilities between the United States and its NATO and other allies.

Developing their own military space applications becomes a concern, lest the military gap grow even larger in terms of modern combat capabilities. Both the Japanese and the Europeans have embarked on such space activities. The Japanese are driven by their growing concerns about North Korea which possesses both missiles and nuclear weapons. Whether the North Korean nuclear weapons are suitable to serve as missile warheads is unclear although both the Soviets and the Americans had to confront that technical problem. In order to placate possible domestic criticism, Japanese efforts are often still cloaked in the rhetoric of commerce and science, but the military applications clearly loom in the background. The European Union (EU) and ESA are building the Galileo navigation satellite system. The military component is clearly embedded in the program – a feature that strongly attracts Chinese interest since it both assists their commercial efforts and enhances their military's efficiency and effectiveness. China lacks the resources currently to create its own global system (although a regional system is being developed, the Beidou), given other priorities, but it has become a participant in Galileo due to European interest in such a relationship. Both are united in their disdain for American efforts to sustain sole control through its GPS navigation satellite system. As will be discussed, China has pursued a regional navigation satellite system, the Beidou, with three satellites in the system.[9]

China straddles the boundary between first- and second-wave national space participants. This reflects its aspirations and its fears. Like the Americans and Soviets, the Chinese viewed space in the form of ballistic missiles as an obvious goal. Missiles came first but space applications were also pursued due to their dual-use implications. But, like the second-wave participants, China also saw benefits flowing (albeit long-term in nature) in terms of economic development. Their view has become more sophisticated over time but the economic imperative meant that space activities had to have a return for the society. The latter stream of activity was obscured at first due to the military-prestige agenda of the first-generation Communist Party leadership. Once the world situation "normalized," China continued both activities but over time pursued space applications for developmental purposes more aggressively.

Third-wave states entered the space arena starting in the late 1970s, a field already populated by several different states operating at differing levels of participation in space activities. ESA fosters participation by states otherwise likely excluded for reasons of limited economic and technological resources. States such as Belgium or Denmark are able to participate, albeit at a lower level than France or Germany, rather then being totally excluded from the field.[10] Their policy decisions were in some ways even more unconstrained in that the earlier participants' multiple successes and failures made clearer the areas of maximum potential gain and greatest difficulty for any future space participants.

The technological advances made by the original space participants, combined with the increasing globalization of the space technology marketplace, meant that the huge start-up costs of the earliest years were not as oppressive as previously, unless the state desired to establish independent space launch (which had dual-use implications). That facet remains prohibitively expensive and difficult. This did not mean that there were no costs, just that those costs were considered manageable depending on the state's space and economic ambitions. As a result, commercial space activities along with scientific applications dominate these newer participants' agendas, in part because military space activities were not considered immediately useful for them and human spaceflight was much too costly for the practical benefits received.

Third-wave states in several cases had initiated space-related activities prior to the late 1970s and early 1980s, but their efforts remained somewhat tentative and unfocused. However, by the mid-1980s, most participants had developed clear national goals for their future space activities, although those goals were always subject to change as events occurred internationally and domestically. India's space program, for example, always struggled for resources in a society with massive social and economic deficits. The Indian space program at first relied upon Soviet launch vehicles, not orbiting independently until four years after its first satellite in 1975. Most such programs therefore were clearly oriented toward providing benefits for the society's domestic needs. These space programs also labor under internationally imposed restrictions because of the growing concern expressed by the first- and second-wave space participants that certain technologies, notably missile launch technology, were being disseminated to potentially dangerous states.

The clearest statement of that cartel arrangement was the Missile Technology Control Regime (MTCR). The MTCR, established in 1987, places severe restrictions upon the dissemination of rocket technologies in the interests of arms control. The official statement is as follows:

> The MTCR rests on adherence to common export policy guidelines (the MTCR Guidelines) applied to an integral common list of controlled items (the MTCR Equipment, Software and Technology Annex). All MTCR decisions are taken by consensus, and MTCR partners regularly exchange information about relevant national export licensing issues.[11]

The membership grew from seven members (the industrialized West) to thirty-four, not including China which has never joined since export of missile technology has been a revenue source. More practically speaking, China was not directly impacted by those restrictions because its program had advanced forward enough not to be totally dependent upon others' technologies, although some technologies were obtained and copied with modifications. India and launch-aspirant states found the MTCR inhibiting – embargoes were imposed, leaving India to develop a totally independent capability if it was to advance. The United States was the political engine behind the MTCR. Many states have found the US position to be hypocritical at times and motivated, not by security concerns, but by the desire to remain economically dominant.[12]

Human spaceflight, on the other hand, is more complicated because no state to this point in time needs to be involved in those activities. Military, scientific and commercial space activities are presently very well accomplished employing robotic spacecraft. This reality becomes even stronger as computers with their associated software grow ever more sophisticated and flexible, making a human presence superfluous. The justification therefore must be more explicitly political in nature in order to rationalize the high costs demanded by human space activities. In order for humans to travel safely to Earth orbit and back, elaborate technologies for their survival against intense cold, radiation, lack of oxygen, and meteorite strikes must be developed. The margins of survival are thin as was demonstrated on July 6, 1971 when three Soviet cosmonauts died when their Soyuz lost air integrity during reentry, and in 2003 when six US astronauts along with an Israeli astronaut died when the space shuttle Columbia's leading edge on its left wing was breached. The Columbia broke up during reentry into the atmosphere.[13]

Political considerations must significantly outweigh the economic and scientific benefits since the latter can be acquired through robotic means. In Table 2.1, the human spaceflight aspect is subdivided into two separate tracks. Only three states to this point, the Russian Federation, the United States, and, most recently, China, operate their own independent launch program for human spaceflight purposes. These states are willing to bear the heavy costs. Any other human spaceflight programs participate in international programs with these three states. China has not (as of 2006) flown individuals from other states to orbit yet, although that would be a logical extension of their program. Thus, states such as Canada, the Europeans and Japan plus assorted other nationals have flown to Earth orbit and beyond multiple times as part of their joint programs with the Russians and Americans. The most recent individual was a Brazilian, Marcos Pontes, who went to the ISS as part of a Soyuz change-out mission.[14] Soyuz spacecraft serve as the lifeboat for the ISS crew but must be changed out every six months. Since there are only two crew, that leaves an empty seat that can be filled with a national astronaut from another state or a rich space tourist.

The governments of these states made the conscious decision that such activities were important enough to pursue at some level but drew the line at implementing their own expensive independent launch capability. The Europeans and Japan, for example, have each pursued independent human spaceflight options but those programs are either temporarily in abeyance or hindered by budget uncertainties. Ironically, American difficulties with their shuttle program after the February 1, 2003 Columbia accident are forcing both parties to consider accelerating their independent efforts. The Europeans had earlier shut down their Hermes space plane but now that the Ariane 5 is operational, their interest may be rekindled through development of the Crew Return Vehicle (CRV). The CRV is conceived as a lifeboat but could in principle be expanded for larger use.[15] The United States had cancelled its portion but the Europeans have found a possible window of opportunity.

Japan's financial issues have blunted their nascent independent crew launch program although technical tests have been conducted on a hypersonic vehicle. The downside of dependence is that one becomes a prisoner of the other state's choices. The United States has a well-deserved reputation for unilateral decisions with only after-the-fact consultations with its supposed partners such as the International Space Station (ISS). However, the space shuttle stand-downs after the two shuttle mishaps would have stranded their personnel on the ground if other arrangements had not been possible.

There are other states for whom cooperative human spaceflight activities become their only realistic options. Their national space budgets are far too small, or other factors make an independent human spaceflight effort politically impossible. Being able to participate in human spaceflight is proving a more fluid choice than originally, if suitable and willing partners exist. Israel sent an astronaut, Ilan Ramon, out on the ill-fated Columbia, a symbol of that fluidity. Israel is unlikely to develop its own space launch for humans; there are too many other pressing economic and military needs to be addressed. Plus, space launch could not occur from Israeli territory given its relationship with its neighbors. Human spaceflight remains the crowning feat for all states regardless of its immediate economic and technological usefulness.

A typology of national space participation

This section introduces and expands the hierarchy of national space participation which underlies the book and provides an historical context for China's program. What makes this hierarchy distinct from earlier overviews of space development is the perceived fluidity of a state's ranking in the hierarchy. From the early space age after World War II onward, there was an assumption or presumption of sorts which argued that a nation's level of space participation was permanently fixed by its economic developmental status. That, as we will see, is more fungible than earlier thought.

Proactive space participation

At the lowest levels in the space developmental process, economic development or lack thereof to a large extent fixes one's position. At the upper levels of economic development, a state's intensity of space activity plainly reflects national leadership commitment. For example, until 2004, the United States was steadily adjusting its expectations downward as regards civil space, with unknown implications for its commercial sector given its tradition of government leadership. This shift in national priorities as to what is considered acceptable reminds one of the field's volatility. Space policy in the United States over its sixty-year history has undergone severe swings in levels of public and elite support for its civil space program. Breaking a space program's momentum can occur quickly, but once forward momentum has been lost recovering may be extremely problematic. Dismantling a space program can be done swiftly; restoring that effort may prove a great deal harder, while one's competitors move forward.

For the United States, civil space activities had wandered to the margins of American national priorities until the Columbia accident in February 2003. Prior to that tragic event, NASA had drifted in terms of programmatic vigor and momentum. In January 2004, President Bush set a new course, a program initiative that encountered general reluctance in the US Congress to support an invigorated civil space program. The political momentum lost earlier is only with great difficulty being recovered in a rapidly changing political environment.

In fact, there is evidence that momentum is not being recaptured given the forces of programmatic inertia and congressional skepticism. The former refers to the legacy programs such as the ISS and Hubble Space Telescope which require support and have entrenched political constituencies. Congress is responding in a mixed fashion, supporting the legacy programs including space science while reluctantly providing additional resources.[16] The agency's credibility is limited given various events, plus external events (the war on terrorism and the growing federal deficit) undermine congressional support.

The greatest fluidity in program composition and purpose occurs among societies in the middle range of economic and technological development. For those societies, space-related activities are not clearly out of reach; the issue is essentially determining at what level a specific society desires to participate in space activities. The answer may be "not at all" despite availability of resources, both budgetary and technical personnel. Germany and the United Kingdom, for example, have scaled back their potential national space activities at different points, ultimately subsuming their major activities under ESA. The respective motivations were different but reflected political judgments. Germany at first felt constrained by its links back to the Nazis and the V-2 rockets, while the United Kingdom's, choices were driven by budgetary issues. Both states participated but, on purely economic criteria, could have done more if the political justifications had been stronger.

That political judgment becomes a difficult one, reflecting shifting social priorities and budget capabilities. Budget capabilities were once seen as fairly fixed in the short term, absolutely determined by a state's level of economic development. That relationship still holds generally but in and of itself is not strictly determinative of policy outcomes. External threats and perceived economic opportunities impact the state's political motivations which then affect their desire to participate. The stronger the perceived threat or opportunity, the more committed national leaderships become in terms of expanding available fiscal capabilities. The former, an external security threat, is the stronger motivation, largely because that is in principle less ambiguous. States can make mistakes and overinflate security threats but that often becomes clear only after the fact. States can usually authoritatively identify their likely foes, although those shift over time.

Opportunities regarding space activities, on the other hand, often come cloaked in problems and uncertainties. The first problem encountered is space launch and how each state is going to handle that question: to develop an independent launch capability or not? Obviously, for some states, there exists no choice since the costs and technologies, especially the former, lie beyond their reach. For advanced states, doing so successfully puts a state on a different policy trajectory from a state willing to remain dependent on others for space launch. Within the space launch arena, there are also fine divisions over what type and size of launcher to build.

Once the decision has been made to build an independent launch capability, the question becomes which orbital location is most useful. LEO is the minimum necessary and for many military purposes and civil applications is perfectly adequate. The next question is how large is the payload capacity of the vehicle? The larger the payload launched, the more useful the likely applications that can be lifted to orbit. Reaching the orbital arc is the most commercially valuable at this point but requires the most significant launch investment. Reaching LEO is the beginning in terms of both rocket technology and applications. Research can be done using sounding rockets but even there the shortness of the flight limits its usefulness. China and India initiated their early space programs using sounding rockets; moving to the next level was what put them apart from other less developed states.

In principle, a more powerful launch system, if created in a timely fashion, opens up other doors. Generating sufficient political support to overcome any domestic opponents takes energy and commitment by leaders who are uncertain as to their priorities or have strongly conflicting preferences. The weaker the state's economy, the more potent social and economic arguments become for investing the state's scarce resources elsewhere. The problem is that the likely rewards or returns from space activity are uncertain in the short term. There are no "sure things" or guarantees in outer space. The prospect of early failures discourages elite engagement in supporting space activities. Politicians possess limited political capital, which they do not want to squander on likely failures. Unfortunately, political hesitation can guarantee failure in a situation otherwise considered promising. By hesitation, we mean the political unwillingness to commit adequate resources to achieve success.

Some governments, such as the United Kingdom, invested sufficient resources to make the initial effort but lacked the political will to continue if failures occurred, and failures will occur if a state is pushing the limits of its knowledge. Much of the messiness in conducting early space activities for the first two space powers was obscured by secrecy at the time. Failure rates in the early US launch program totaled nearly 50 percent and almost led to a shutdown unless the technical issues were resolved. These problems were in time overcome through developing new approaches to building and implementing the technologies. The Soviets encountered similar issues but secrecy kept many out of public view at the time. Both programs, for example, lost astronauts and cosmonauts.

For new aspiring participants, failure occurs in several ways. First, the space initiative is just plain underfunded or unsupported against attacks by its opponents who value their particular priorities more highly. Second, failures can occur because the space program is funded erratically or sporadically, destroying any momentum or continuity. The US space effort has several times endured stretch-outs when critical funding decision points are delayed as to implementation or the promised peak funding is not allocated, thus effectively negating much of the work done up to that point.

Teams of highly skilled personnel are broken up and dispersed to other projects, and reassembling even a few takes precious time. The space shuttle development process was stretched out over years, as was the International Space Station (ISS). The reasons for this reluctance are multiple but the clear lack of interest among much of the US federal government for supporting the shuttle or ISS was crucial.

In the case of China, fluctuations in resources for space activities obviously reflected changes in its leadership's priorities or, more usually, periods of transition in power from one group to another. Fortunately, for space adherents, Chinese space activities have fallen under the general rubric of national security activities, which has reduced the fluctuations experienced compared to a purely civil space program. Linkages to China's military have been important because, as indicated earlier, the leadership feels a strong sense of external threat from other states, especially the United States. This sense of threat fluctuates over time in its intensity but has never disappeared. The more usual problem, politically speaking, remains the demands for repeated large additional infusions of cash in pursuit of new or upgraded space technologies. All leaderships eventually balk at these new requests, especially when the benefits are not considered commensurate with the costs. China in particular was prone to such behavior given the political turmoil and harsh budget realities confronting the state. Overall, the perceived benefits were thought to outweigh the costs.

Regardless of these problems, individual states engage in very different levels of space participation. The particular levels reflect, in differing combinations, their history, level of economic and technological capability, threat perception, and national aspiration level. The intensity of any particular nation's space effort becomes the product of their motivation to participate and the society's political capacity to divert scarce resources to this particular activity. Societies lacking in effect the requisite economic surplus must make very hard political choices. Governments that are authoritarian in nature can at least in the short term ignore possible popular discontents, but no society over the long run has proven immune to internal unrest if critical social and economic needs are not met. The Soviet Union collapsed for this reason among others.

As will be discussed in subsequent chapters, China chose to pursue a first-rate space program, including human spaceflight, while selectively implementing those technologies necessary to support its aspirations. The Chinese space program has focused on those essential technologies while eschewing a comprehensive program. This approach was considered the only feasible way to afford a space program commensurate with Chinese international aspirations. The areas chosen are all ones where China could in effect bootstrap itself to the requisite level necessary for achieving success. The footsteps of the early space pioneers provided them with the necessary directions as to how to proceed but also allowed China to move forward independently. Becoming a hostage to another state was the one option totally rejected by China.

In the past, national engagement in space activities normally presumed the participant country operated at the leading edge. If a state was thought "behind" that normally implied obsolescence and ultimately failure in terms of succeeding regarding space activities. In fact, the speed of recent space-related application development has meant that the proliferation of profitable commercial opportunities occurs at less than cutting edge. Likewise, military applications need not be cutting edge in order to significantly enhance a state's power. The particular needs of the users (whether government or private) are fully met at that lower level. More sophisticated technology may prove faster and possibly cheaper on a per volume basis but the reality is that such a level of sophistication is not immediately necessary and might not meet the needs of the user state. One must remember that engineers have a professional bias toward maximizing technological innovation even when that is not necessary for achieving the desired goals.

It is suggested that the hierarchy of space participation (see Table 2.2) represents a declining level of engagement by states as one moves from Level 1A to Level 4. Briefly summarized, there are four levels specified within the suggested hierarchy. At the top, under "Space faring participants," are those few states traditionally associated with being heavily engaged down to those nations whose participation is more likely based on objective physical features, usually their global location. These states represent at a crude level space infrastructure in its most primitive sense.

The first and most intense level is delineated by whether the specific state has developed an independent spaceflight capability. This capability is a more developed one than merely demonstration activities where military missiles are crudely modified in order to achieve LEO. The separation marker here becomes achieving sustained spaceflight capability. Though vague, this conception embodies the idea that the involved party envisions and achieves "routine" access to space based upon employing their own space

Table 2.2 Hierarchy of space participation

Level	Type of participation	Example states
Level 1	Space faring participants	
A	Crewed space flight	Russia, US, China
B	Spaceflight	Japan, Europe, India
C	Newly emergent	Israel, Brazil
Level 2	General space participants	Germany, UK, France, Italy
Level 3	Selective space participants	Belgium, Norway, Canada
Level 4	Passive space participants	Tonga, Kazakhstan

assets. Within this larger category of flight-capable, there exist clear distinctions indicating differences in level of activity and associated intensity.

Under the general rubric of "space faring," there exist three distinct levels, with crewed spaceflight being the highest level of effort. This reflects the fact that any launch system must be "human or crew rated," meaning safe enough for humans to fly on the vehicle. Crew rating can be achieved through provision of safety systems or some means of crew extraction during liftoff, or rescue when in orbit. The latter was not a particularly salient consideration previously but the Columbia accident in 2003 raised the question: what would have been done if NASA had been fully aware of the wing damage. How do you rescue astronauts stranded in orbit and unable to reach the ISS for refuge? The answer was considered to be nothing really, although after the fact scenarios were discussed.

Achieving this safety level demands significant resources and always verges on the edge of catastrophe especially during liftoff when vast amounts of energy are released quickly. Only three states have achieved this status – one that may prove difficult to sustain without political support. Few people have died during spaceflight but this only reflects the fact that very few have ever flown to orbit. That number will increase, but how fast depends on political and economic variables that are continually changing. For example, in the aftermath of the SpaceShipOne suborbital flight in October 2004 which won the X-Prize, there was much talk about dramatic increases in space tourism.[17] Whether those numbers will in fact increase, however, depends upon the industry developing reliable and safe lifters with rapid turnaround times – the holy grail of space launch for several generations.

"Spaceflight" and "newly emergent" reflect national programs which possess and operate independent space launch capabilities albeit at different levels. The former refers to states operating at all levels of space launch below crewed spaceflight, while newly emergent refers to states working on or having just attained LEO capability. All these states have made the decision to develop such flight technologies – the measure *par excellence* of engagement in space activities at a serious level. For China, the process of moving from newly emergent in the early 1970s to successful human spaceflight took nearly forty years, reflecting the challenges to be overcome.

The next level down, Level 2, encompasses states who are not routinely or independently flight-capable but whose involvement is general otherwise. That is, these states have made a definite commitment to participate in space activities but not in the launch arena. Their choice can reflect economic judgments or political decisions based upon national history. Germany, as mentioned earlier, chose earlier not to become flight-capable, reflecting its historical legacy from World War II. This reluctance has been fading since German reunification after the fall of the Berlin Wall, and preliminary work on rocket technology has begun again, the first step toward achieving independent flight capability. Germany earlier (as West Germany alone) had the technical and fiscal ability to participate; the deliberate choice was not to do so.

For European states generally, the European Space Agency (ESA) has substituted for independent national spaceflight and applications. All ESA member states have the opportunity for independent activities outside the purview of ESA, but, for all, that does not presently involve independent spaceflight. Within the context of the hierarchy, ESA ranks collectively as a Level 1B or spaceflight participant while individually the member states fall on Levels 2 and 3. Level 2 (general space participants) includes states such as France and Italy, both of whom operate large national space efforts in parallel with ESA.

The bottom two levels represent a significant decrease in national involvement. Level 3 or selective participants see themselves as forced to leverage their assets by cooperating in multi-nation endeavors. Selective participants include the smaller European ESA members along with Canada and Singapore. Their interest is often in principle economically and technologically defensive in that the desire is to keep in touch with the field. These states more often include purchasers of services rather than providers, although some such as Singapore saw great profit in value-added services such as processing remote sensing imagery for commercial use. Space activity is seen as another way by which to leverage their limited resources by tapping into data acquired by others. Their participation levels are likely erratic in that shifts in national priorities can drastically alter or hinder their efforts. In fact, some states can recede almost out of the picture, depending upon national priorities. Australia, for example, was involved in early space activities, primarily as a launch facility. When the first European launch project, Europa, failed, Australia basically withdrew from the field. More recently, several proposals have been advanced considering Woomera in Australia as a potential spaceport location, and possibly Cape York. Those efforts have foundered up to this point due to market changes in commercial space activities but the federal government has come to support space technology development, at least selectively. Australia has therefore moved up from Level 4 to Level 3 due to decisions made linking it back to the larger space community.

The lowest level of participation, Level 4, is termed passive only because the different states become involved due to some national physical feature, usually involving location. Space activity in principle is beyond their present capabilities so their involvement becomes indirect through the activities of more sophisticated participants, such as the involvement of Kazakhstan with its Baikonur Cosmodome. An obvious example would be as the site for a spaceport, in which location – probably close to the Equator – is essential, but national involvement otherwise is minimal. The location can also be orbital: for example, the Kingdom of Tonga has leased out its orbital slots over the Pacific Ocean – useful locations for comsats. There are economic benefits derived from that activity but at a lower level than those of the operators. Spillover effects into the general economy are normally assumed to occur but are often built into the relationship.

Conclusion

China, over the past forty years of its space activities, has through its efforts advanced from the lowest level to the highest. Assistance from other states, especially the Soviet Union, originally helped but political differences ended that cooperation for a decade. The development process is and was not easy especially for a state confronting severe economic and political challenges. Its program was consciously accelerated because China confronted external security threats, threats that China found very credible.

China's space program is a hybrid in that its motivations are a combination of first wave (security and prestige) and second wave (economic and social development). It is this fluidity of purpose that strikes one as China grapples with managing a burgeoning program whose purposes arouse either fear or envy (the security question) and economic competitive concerns. When one surveys the hierarchy of space activities, the uniqueness of what China has accomplished becomes clearer.

Chapter 3

First awakenings

Introduction

Even before the full blossoming of the space age in the 1950s, China's leaders were very interested in joining that effort. Mao and other Chinese Communist leaders had witnessed World War II and the Chinese civil war and were especially aware of the military gap between China and the West. The roar of imported American artillery used by the Kuomintang army and the explosions of the atomic bomb in Japan made China fully cognizant of modern military technology's power. Their interest was especially piqued by the prospect of China's possession of ballistic missiles capable of reaching targets, or later orbits. The former aspect was more immediately compelling than the latter. Developing missile technology quickly was closely related to the hostile international environment. China confronted the superpowers, first the United States and, later, the Soviet Union.

China's experience included the fact that at the end of the Korean War President Dwight Eisenhower directly and secretly threatened China with the possibility of nuclear attack if a truce in Korea was not established. China clearly lacked any credible deterrent to the powerful American nuclear bomber forces, the B-29 first, then the B-36, later followed by the B-47 and B-52. For all intents and purposes, China lay totally open to its enemy. The United States was aggressively over-flying Chinese air space searching for elements of their air defenses and interceptors. Such violations also occurred with regard to the Soviet Union but the Soviets possessed some retaliation and defensive capabilities and acquired more when their rocket forces matured in the late 1950s. The international crisis over the 1960 shooting down of Gary Powers in his U-2 spy plane signaled the Soviet self-defense capability to the West. Thus it is not difficult to understand that the early Chinese space programs were strongly military oriented. That military foundation was similar to the Soviet and US programs which grew out of their missile programs. Reaching orbit became possible much earlier than otherwise likely due to these military priorities. Outer space was not their goal except as a medium through which missiles would travel toward their targets.

Missile programs developed quickly in the 1960s; by October 1971, China could launch an ICBM, which enhanced China's capacity for strategic nuclear deterrence, the first priority at that time. Other space activities followed rather than led their efforts. This priority was duplicated in the other states. President Eisenhower approved the first US space launch effort only on the condition that the efforts did not interfere with the ICBM development program. Reaching the Soviet Union was clearly the highest priority rather than reaching outer space. In fact, Wernher von Braun argued that Eisenhower's directive stopped the United States from reaching orbit a year prior to the Soviets. True or not, the focus of missile programs was clearly not space access but attacking one's enemies. A pattern found in all three programs – Soviet, US and Chinese – space was a sideshow until after the launch of Sputnik and the emergence of the great power space race.[1] For China, the effort was far more challenging; overcoming those challenges became the work of the first generation, making dreams reality.

Its issues of economic underdevelopment and political isolation made China's efforts more challenging than successful at first. The political schism with the Soviet Union, starting in August 1960, further exacerbated their problem by cutting off China's only immediate source of technical support. China was unique among Soviet allies because it was more independent politically and economically – a fact which was exacerbated by personality conflicts between Nikita Khrushchev and Mao Zedong. Their visions of the immediate situation were conflicting, especially over the question of US strength and resolve. The Soviets essentially cut ties because of their more cautious view of the world. For China, the Soviet departure disrupted its technological progress and threw the Chinese back on their own resources. However, the Soviets had provided some technical assistance, especially in training, which could be put to use, although the Soviets were very reluctant to reveal much.

Space technologies, if domestically produced, demand a certain high level of technology development plus the requisite economic surplus to invest in the program. At first China was not capable of sustaining either of those requirements, which caused a series of frustrations and launch failures of prospective missiles. China in the 1950s was extremely backward in science and technology. Wars, external and civil, had lasted for almost a hundred years before the Communist Party unified mainland China in 1949. China's nascent technological infrastructures were severely devastated and qualified personnel in science and technology were scarce. Thus, unlike the Soviet Union and the United States, China completely lacked the necessary technology and requisite economic resources to build a space industry. But strong support from the state eventually overcame these constraints and an active space program was created, although the space program always labored under the burden of China's economic problems – meaning that its resources were always subject to attack by groups arguing their greater priority for meeting other national needs.

Chapters 3 and 4 explore the major themes that defined Chinese space policy originally and over time. The analysis here draws upon earlier work by Yanping Chen and the official statement of the Chinese Academy of Space Technology which conceptualized the development of Chinese space activities as falling into four time periods.[2] The first era from 1956 until 1966 saw the establishment of the space program and its early steps. The second period spanned the period of the Cultural Revolution, 1966–1976, ending with the death of Mao Zedong. These two periods will be briefly described and analyzed in this chapter. Chapter 4 covers the latter two periods.[3] The third period spanned 1976–1986 when the space program not only survived the competing priorities but also evolved into a coherent and growing program. The last period, from 1986 to the present, finds the space effort an integral part of national research and technology policy plus a growing source of Chinese pride and nationalism, a tool for furthering nation building.

The first era (1956–1966)

First things first – space programs for security construction

China was aware from reports in the open literature including the press that the Soviet Union and the United States were moving forward in pursuit of intercontinental ballistic missile (ICBM) force development. The Americans had their Atlas and Titan missiles well underway, both capable of reaching across the globe. In addition, development of intermediate range missiles and submarine-based missiles were other options being pursued. In the 1950s and early 1960s, there were open discussions about tactical nuclear warfare and escalation strategies, all of which were perceived as threatening to China. The Soviet missile program was much more secret but evidence existed through Western media and Soviet sources of their equivalent missile program. In fact, American fears of purported Soviet missile strength fueled talk of a "missile gap" on the Americans' part, fueling the next round of the nuclear arms race. American fears were fueled by the lack of specific knowledge about Soviet intentions and military assets. This made space-based reconnaissance a priority once Earth orbit was attained.[4] Mao Zedong was already suspicious of the Soviet leadership – seeing them as a threat to Chinese state autonomy and less ideologically driven than China, which he defined as a revolutionary state. More critically, China was militarily weak, incapable of successfully defending itself or deterring potential nuclear attacks. Chinese political rhetoric spoke of paper tigers in reference to the United States but there was a clear awareness of its nuclear teeth. Building an equivalent Chinese nuclear bomber force against the United States was not feasible, and the Soviets were of marginal usefulness since their bomber forces remained inferior to the American Strategic Air Command in the 1950s. How to overcome this shortcoming became China's overriding military priority.

To achieve strategic parity if not superiority against the United States, the Soviets arrived at the conclusion that the shortcut lay with accelerated development of their ICBM forces. Therefore, their primary focus became developing larger and more accurate military rockets. Payload capabilities had to be raised because initially any nuclear warheads were large and heavy. That problem had even impacted the ability of bombers such as the American B-29 to carry the H-bomb. The Soviet approach became one of brute force in order to lift and deliver nuclear warheads. The Americans by contrast went for miniaturization rather than more powerful lifters. The result was that during the space race, Soviet payloads were always significantly larger and heavier than their American counterparts, until their Saturn 1B and Saturn 5 vehicles flew as part of the Apollo program.

China lagged behind both states and its decisions as to how to proceed were conditioned by the available and possible resources, which included seeking access to the others' rocket technologies. The first Chinese attempt to acquire Soviet rocket technology in 1956 resulted in delivery of two R-1 missiles. These were simply Soviet copies of German V-2 rockets. Tsien Hsue-shen, who had been to Germany as part of the American program aimed at acquiring and evaluating Nazi rocket technologies, led the Chinese effort and was unimpressed by the Soviet-supplied materials. The R-1 rockets did not help China advance the state of their art. After further negotiations, the Soviets agreed to provide an R-2 missile along with the relevant plans – a step forward in terms of the level of missile technology involved but not equivalent to where the Soviets were technologically speaking at that time. That reflected both superpowers' reluctance to assist potential rivals even though presently allies.

Tsien Hsue-shen had returned from the United States after teaching and researching at Cal Tech and the Jet Propulsion Lab. His return to China came after years of path-breaking work in the early US rocket program. Tsien's presence, along with whatever assistance was provided by the Soviets, has fueled continuing speculation that China's space program is merely a copy of others' work. Obviously, Tsien's knowledge did not disappear, but his arrival in China was delayed for years, allowing his knowledge to become outdated. Soviet assistance was both grudging and limited, especially as political disputes grew more heated. China's space program tracks the pathways of the first two programs, borrowing when useful and possible, because they had the relevant experiences which China lacked. China's technological backwardness, however, meant that in the end there were no shortcuts – indigenous resources and personnel had to do the work.

China clearly was not a significant player in the early space age. In fact, China's original efforts at developing an independent launch capability failed miserably. The failure reflected its dearth of basic technology capable of producing the equipment able to endure space launch.

At first, Chinese technologists were given limited access to Soviet training and technologies. This initial access has long been cited as indicating that the Chinese space program is simply a pale carbon copy of earlier Soviet technologies. Similar comments were made regarding the Soviets at first – that their technologies were all derivatives of German technologies rather than products of their independent technology programs. What that analysis ignores is the bitter August 1960 split between China and the Soviet Union – the split was a volatile mixture of ideology, policy and personality. The ideological dimension and the policy aspect merged, with the Chinese proposing an aggressive confrontation strategy versus the United States while the Soviets professed similar sentiments but their strategy was more pragmatic and less confrontational. Mao saw nuclear war as much less frightening than the Soviets viewed it – China's vast population would allow its regime to survive. The Soviets were becoming less sanguine about that possibility. From their perspective, the Chinese had less to lose than the Soviets. Nikita Khrushchev and Mao Zedong as personalities exacerbated the split, but, for China, the immediate impact was to completely sever its connection to the Soviet space and missile program, a connection that had proven bigger on promises than delivery.

For China, this abrupt rupture had two immediate effects: (1) progress was slowed while the missile program reoriented itself, and (2) it forced China to develop its own indigenous technologies. Technology development dramatically slowed because problems now had to be solved with minimal outside assistance. There were no shortcuts to space for the Chinese. Much of the outside assistance received came from the open technical literature from the West. This reliance on the open literature, especially that produced by NASA, had profound effects upon the program. It did not provide China with space launch capability, which still had to be developed. China had a small cadre of US-trained engineers (returned from the United States where they had gone earlier as students) while others had acquired some Soviet training. The Soviet influences did not immediately or ever completely disappear since the initial direction of the Chinese program had been set in those early years when the Soviets were somewhat helpful and clearly present. China was forced to rely upon its own resources if progress was to be achieved.

China at that time was an economically backward country, with an estimated output of electric power of 12.5 billion kilowatt-hours and steel production of only 4.12 million tonnes in 1955.[5] The bottlenecks of economic difficulty and technological backwardness could not have been overcome without strong political support. Scholars studying developing countries argue that state autonomy is necessary for development.[6] This is particularly true for the development of space technology, since private sectors responding to the market cannot make sufficient investment in R&D in this field. China possessed high state autonomy and state capacity with its planned economy and the state-controlled necessary resources to achieve its goals. The development of space programs was put in a prestigious position directly under the leadership of the party chief and government head and clearly benefited from the centralized political system and the largesse of the state. In March 1956, the State Council passed the *Long-Term Plans for Scientific and Technological Development, 1956–1967.* Missile technology was included as a major national priority with other embryonic space programs centralized under the direct leadership of the Central Committee headed by Mao Zedong. Although the specific bureaucratic agency designated in charge of space programs was first the State Aeronautics Industry Commission and then the Seventh Ministry, they only carried out the decisions approved by the Central Committee. The agencies could not even independently establish their subordinate R&D institutes, reflecting the centralized style of Chinese S&T organization. In 1956, for example, the head of the State Aeronautics Industry Commission had to get Zhou Enlai's approval in order to establish the R&D institute for missile defense. Specific space programs, whether missile programs or satellite construction, had to obtain final approval from the Central Committee chaired by Mao Zedong.

The program's salient position in the power hierarchy maximized the probability of sufficient resources flowing to space programs under the state-planned economy. After DF-1 (the Chinese version of the R-2) made its maiden flight in 1960, the government allocated more resources for developing the DF-2. DF-2 was the first indigenously designed missile and was successfully tested in June 1964. Soon DF-2 was redesigned as DF-2A to carry an atomic bomb, a capability China acquired in October 1964 with their first nuclear test; the DF-2A was tested in October 1966 with the atomic bomb exploded in the targeted Xianjang desert a thousand miles away. This test was startling to outside observers (and also a warning) because of its use of a missile to deliver the weapon at its test stage. Western states and the Soviet Union had generally employed static weapon testing methods in order to minimize possible mistakes with horrendous results. The possibility for accidents was always critical when considering nuclear weapons. R&D on the DF-3, the intermediate range ballistic missile capable of reaching the Philippines, started in 1964 and its first successful test flight was made in 1966. China experienced growing success in its technology developments, which came despite severe economic crises and political instability within the regime itself.

China's economic situation was dire in the 1950s and the 1960s. In the three famine years (1959–1961), the number of people who died from malnutrition or from other diseases caused by food shortages was estimated at between 15 and 30 million.[7] The satellite project, which started in 1958, was suspended in 1959 due to insufficient resources, but the missile program was never stopped. This was but another piece of evidence demonstrating the overwhelming security focus of the program. In this, China mirrored (albeit on a smaller scale) the path taken by the two space powers. After a series of failures, the DF-1 missile made its first successful flight in 1960, while millions of people in the rural areas died of hunger. Why were missile programs fervently supported? The answer lies in the mounting external pressures, first from the United States and later from the Soviet Union. The regime's sense of isolation and threat was accentuated by the superpowers' nuclear arms race. That race further demonstrated China's comparative weakness. China had no nuclear weapons while they had thousands.

From 1949 until the early 1970s, the United States and China were in a completely hostile relationship. To the United States, China was an aggressive, irrational and unpredictable power,[8] while to China, the United States was an evil force supporting Chiang Kai-shek and preventing mainland China from unifying China. As early as February 1950, the People's Liberation Army (PLA) was trained to cross the Straits to liberate Taiwan.[9] This plan was postponed first by the Korean War and later by the "nuclear blackmail" of the United States. The Korean War, two Taiwan Straits crises, and the Vietnam War all accentuated the mutual resentments based on ideological and political disputes. In a deep sense, China, after a century and a half of European semi-colonialism, was reacquiring what its leadership perceived as its traditional role in East Asia as a dominant power. These efforts, however, ran up directly against US containment efforts – China being considered an integral part of the Soviet bloc.

After the Korean War broke out in June 1950, the United States saw Taiwan as a competitor to the mainland government and thus assisted Taiwan with military support. In fact, American policy makers prior to the North Korean invasion had provided only minimal support for the Nationalist government in Taiwan, especially regarding any return to the mainland. After the invasion, however, the United States at least rhetorically reversed course and extended its nuclear protection to include Taiwan.[10] In fact, the invasion is considered to have occurred only because Secretary of State Dean Acheson in a speech at the National Press Club verbally drew a line excluding South Korea. Joseph Stalin supported the initial incursion but Chinese forces intervened in October 1950 after the North Korean army was in retreat. The Chinese forces were "volunteers" which reduced the potential for a full-scale war with the United States only a year after coming to power.[11]

In February 1951, the United States and Taiwan signed a Mutual Defense Assistance Agreement, in which the United States promised to provide Taiwan with military equipment so that Taiwan could mount a defense against possible attack. The result has been a long-term conflict over the future of Taiwan, with a variety of diplomatic formulae being used to keep conflict low.

In the eyes of most Chinese, the United States not only hindered China from completing unification, but also assisted Chiang Kai-shek in attacking the mainland. Chiang remained a threat to the mainland Chinese government at the beginning of the 1950s and, in fact, he did not give up his hope of recapturing the mainland at that time. In 1954, with the acquiescence of the US government, the Nationalists seized several groups of offshore islands. After they controlled the islands, the Nationalists used them as military bases to harass China's central coast assisted by American planes.[12]

The mainland's attempt to unify Taiwan failed due to American nuclear deterrence. On March 16, 1955, President Eisenhower in a news conference threatened to use tactical atomic weapons in any war against China.[13] When the PLA shelled the island of Quemoy in 1958, the same thing occurred; the military action by the mainland had to stop due to the nuclear threat of the Eisenhower administration.

Another unforgettable event of this era was the Korean War. Facing the advanced weapons of the UN forces headed by the United States, China lost between 300,000 and one million people, including Mao's son.[14] The United States "threatened to widen the war through use of nuclear weapons on the Chinese mainland."[15] Therefore the three nuclear threats from the United States were the most important reason driving the Chinese government to construct its own deterrence force in order to break the "American nuclear monopoly" and to smash "America's blackmail."[16] Together with developing nuclear bombs, the weapon carrier, intercontinental ballistic missiles, became the most urgent developmental priority for bolstering Chinese defense. The ballistic missiles developed in the early era were aimed at American military bases. The first indigenously designed missile, the DF-2, targeted American military bases in Japan, while the DF-3 was designed to strike those located in the Philippines. The DF-4 was intended for Guam and the DF-5 for continental America.

The Sino–Soviet relationship began to deteriorate in 1957 and 1958 due to several reasons, which have been much discussed and thus will not be dwelled upon here. The mutual hostility became completely open in 1963 when shrill diatribes ensued. At this same time, serious border clashes occurred along their mutual borders. According to Chinese statistics, 4,189 border clashes with the Soviet Union were reported between October 1964 and March 1969.[17] Both countries reinforced their troops along their eastern and western borders. After the first DF-3 test was successful in 1966, R&D on a redesigned version was started so that Moscow could be brought within its range. The military threat from the Soviet Union also made China build its second launch center, the Xichang Launch Center. In 1969, after the great split, since Soviet experts might know some of the technical parameters of the Jiuquan Satellite Launch Center (the first launch center in China) from their participation, and because Jiuquan is geographically near to the Soviet Union, China decided to build a second launch center, the Xichang Launch Center, which is located at 102.0 degrees east and 28.2 degrees north, far away from both the Soviet Union and the coast area, which was more vulnerable to possible attack from the Americans'. Thus China's pursuit of nuclear deterrence remained the driving force, invigorating its efforts to both build ballistic missiles and ultimately launch vehicles. The priority was clear, defense first over all other uses.

Space programs and national prestige

When China decided to construct its first satellite in 1965, it was facing an unprecedentedly hostile international environment. From the Chinese perspective, that meant that potential adversaries were not necessarily deterred by mere possession of nuclear weapons if there existed no reliable delivery method. The verbal attacks between China and the Soviet Union became more vitriolic, while border clashes grew. The Vietnam War escalated and Mao perceived it as a signal for war.[18] Although China exploded its first atomic bomb in 1964, the outside world knew little about its missile program. The sounding rocket launched from 1960 to 1965 conducted some biological experiments but those occurred in obscurity and were not militarily impressive. Thus the rationale for launching their first satellite was partially to demonstrate its deterrent ability and thus to enhance China's national prestige. The original proposal for constructing a satellite was approved by Mao in 1958 but was suspended in 1959 due to insufficient resources. In 1965 the satellite project was restarted, with the first satellite launched in 1970. With that success, China became the fifth country in the world to successfully launch a satellite. Chinese even considered a manned space program at this stage but that proved premature for technical and economic reasons. The first manned spacecraft, Shuguang-1, was under serious consideration but political turmoil finally closed this door for a time.

A domestic propaganda tool

The additional gains brought by the space program in domestic politics were that it served as a propaganda tool. The economic situation during this era was dire and Mao certainly needed some successes in order to demonstrate his leadership ability. The space program served as an excellent social stimulant. The planned economy together with China's authoritarian political structure left little room for people to make a choice; psychologically the populace needed the government to provide some incentives, albeit intangible for most at this time, especially in a country without religious belief. In addition, even in authoritarian countries, the paramount leader still requires success that legitimizes their rule. The failure of the Great Leap Forward and the resulting famine, which lasted for three years from 1959 to 1961, clearly weakened people's belief in the leadership capacity of the party; not only Mao but also the collective party leadership needed to demonstrate their leadership ability with public successes. Satellite launching thus became a rational choice of the leadership. The ruling elites even used the space program as a tool to inspire personal worship. The Dong Feng series name, for example, comes from a Mao saying that the "East wind prevails over the West wind": that is, the strength of socialism exceeds the strength of imperialism. Hence it is not difficult to infer that the space program in this era became one of the strategic forts that different political factions fought over.

Space programs and political power enhancement

Successful space programs bring domestic political gains for every country, but in authoritarian countries, controlling the space program can bring more domestic gains for paramount leaders, whose power essentially relies on the military's support. Missiles represent one of the most advanced modern weapons, while other space technologies, such as remote sensing, surveillance, satellite communications and satellite navigation, all play an important role in modern wars. Space programs in authoritarian countries are more likely to become strategic areas over which different political factions fight to acquire control. Different from the American problem of "to be or not to be" (in other words, whether the government supports a certain space program or not), the Chinese problem was "to be but for whom" (i.e. to whom the space program swears its loyalty). Chinese domestic politics were factionalized in the early 1960s. "By late 1963 there were, in effect, two Chinese Communist Parties: One was headed by Liu and Deng Hsiao-ping and operated through the civilian CCP apparatus. The other was the military party system, dominated by Mao and Lin Biao."[19] Deep and continued political involvement by top leaders of the different factions became an important characteristic of the Chinese space programs of this era.

Mao directly and repeatedly intervened in the space program. Even during his semi-retired years (1959–1966), when the space program was not his official responsibility, Mao still kept close watch on its progress. Mao at least twice attended S&T exhibitions related to space program development, important symbolic events in Chinese politics. He even intervened in some trivial decisions. For instance, the specific time of launching DFH-1, China's first satellite, had to be approved by Mao.[20] It was not only Mao who monitored the progress (or failures) of space programs; other political factions did the same. Before 1962, missile programs were under the direct leadership of the Defense Ministry headed by Lin Biao. But in 1965, the Fifth Academy changed from a military institution to a non-military institution designated with a new name – Ministry of Machinery Industry. This dynamics of military versus civilian control very likely meant that Liu, the head of the bureaucracy, was attempting to circumvent the Defense Ministry and impose his direct control over the missile programs. Since Defense Minister Lin Biao was Mao's strong supporter, it seems that Liu and Deng were attempting to weaken Mao's interference by taking the missile programs out of the military domain. This move proved more problematic during the Cultural Revolution when the program came under Red Guard attack, no longer protected by its military connections. The Mao faction's control was temporarily weakened. In the heyday of great power competition in 1965, one year before the Cultural Revolution, different political factions inspected the Ministry of Machinery Industry, including Liu Shaoqi (Chairman of the People's Republic of China and Mao's political rival) and Mao's wife, Jiang Qing (one of the core leaders within the Mao faction).

Economic and technological constraints

Compared to the United States and Soviet Union, China was clearly a newborn in terms of international space programs at this point. By 1965, the United States had made 432 orbital launches, including successful launches of 11 communication satellites, 8 ELINT satellites, 20 meteorology satellites, 18 navigation satellites and 106 reconnaissance satellites. Manned spaceflight, lunar probe and other deep space explorations were also carried out. The Americans fifty years ago did what the Chinese are doing and planning to do currently. The Soviet Union had made 188 orbital launches, including 25 military communication satellites and 46 reconnaissance satellites, and curried out other activities including lunar probes, planetary missions and manned spaceflight. Besides the two space giants, other nations, such as Canada, France, and Italy, had also launched their own satellites although in some cases they had traveled on others' launchers. China's underdeveloped status hinged on two constraints – economic and technological constraints.

Economic constraints

Space program R&D, particularly in the early years, requires huge upfront investments. For example, the Apollo project cost 255 billion dollars from 1961 to 1972. The first fifty orbit launches took the United States less than five years, the Soviet Union six years, Japan twenty-four years and China twenty-eight years. The negative correlation between economic prowess and the time of success indicates the importance of space program investment. In Japan's case, the failure occurred in the first ten years, but after 1975, Japan invested hugely in buying American mature space technology, which dramatically improved their launching success rate.[21] But China obviously lacked enough resources. According to the data provided by the Groningen Growth and Development Centre, the average GDP of China (1956–1965) was 425,389.30 (in millions of 1990 US dollars, converted at Geary Khamis PPPs),[22] one-half of that of the Soviet Union and one-fifth of that of the United States. If we examine GDP per capita, China was significantly poorer. The average GDP per capita of China (1956–1965) was $637.50, one-sixth of that of the Soviet Union and one-eighteenth of that of the United States. Although Mao's regime did not care so much about people's living standards as later regimes, economic constraints still impacted the early Chinese space effort. For example, the R&D on submarine-based missiles, which started from 1958, had to be suspended due to economic difficulty in 1963. Work on the first satellite was another example. The satellite project started in 1958, but was suspended in 1959 due to insufficient resources partially caused by the big famine created by the Great Leap Forward.

It should be noted that the planned economy and authoritarian political structure maximized the probability of Chinese space program development at that time. While between 5 and 30 million Chinese died of hunger and people in rural areas ate nothing but tree leaves or even earth during the three famine years (1959–1961), the missile program technical personnel were allocated adequate rations. These comparatively luxurious rations were guaranteed for every household of the missile team members by a state service agency. The government also invested more than 1 billion yuan to build the missile research program. Besides, the Chinese pattern exhibited higher risk tolerance in their space program development compared to Western countries. It is understood that space investment involves high uncertainties in terms of eventual success. China, as well as the former Soviet Union, did not care as much about adverse public or international opinion as the Western governments did, and thus resources were much easier to aggregate and to invest in high risk projects. This insulation from public opinion but not elite opinion made resource allocation easier compared to the situation in democracies. However, abrupt changes could occur as the result of splits in elite views regarding the space program or, at least, certain aspects of the program.

Technological constraints

China's planned economy could partially redress the economic constraints, but it could by no means redress the problem of technological constraints. Although having led the world for one thousand years in science and technology, China's role declined after the fourteenth century. Isolated from

Table 3.1 Examples of returning scientists and technologists

Name	Specialty	Returned from	Year of return
Chen Fangyun	Electronic engineering and system engineering specialist	UK	1949
Chen Kuanneng	Metal physicist	US	1955
Guo Yonghuai	Aerodynamicist	US	1957
Qian Xuesen	Jet propulsion specialist	US	1955
Ren Xinmin	Space technology and rocket engine specialist	US	1949
Sun Jiadong	Rocket and satellite specialist	Soviet Union	1958
Tu Shoue	Rocket specialist	US	1945
Yang Jiaxi	Automation specialist	US	1956
Yao Tongbin	Spatial metallurgist	UK	1957
Wang Dayan	Optics specialist	UK	1948
Wang Xiji	Recoverable satellite specialist	US	1950
Wang Weilu	Rocket (missile) specialist	UK	1947
Wu Ziliang	Metallurgist	US	1950
Zhao Jiuzhang	Earth physicist	Germany	1938
Qian Ji	Space technology and physicist	NA	NA

the rest of the world deliberately, China lost the chance of remaining up to date with modern science and technology before 1842. After 1842, China gradually accepted modern science and technology and began sending students to study abroad. Although, overall, science and technology in China before 1949 was in the swamp of backwardness, those who returned to China became a spark, igniting Chinese science and technology by devoting themselves either to research or teaching. Among the twenty-three prominent scientists who made a special contribution to the "Two Bombs (atomic and hydrogen bomb) and Satellite Projects," twenty received graduate degree from foreign countries (see Table 3.1).

Although these scientists played an important role in knowledge creation and distribution in the development of the Chinese space programs, it should be noted that in China the foundation of modern science and technology initially remained poor. China was late in entering the field of modern science and technology, which was not supported by the Chinese government until the 1860s (the Self-Strengthening Movement). In 1900, nobody in China knew infinitesimal calculus, but in Western countries the subject was taught as a first-year college course. Systematic training in subjects such as mathematics was not established in China until the 1920s. Moreover, recurring wars lasting for almost 100 years devastated the economy and consequently research funds for R&D were negligible. By 1949, China nominally had about 50,000 trained technical personnel, but only 500 of them were real researchers. Modern science and technology only existed in limited areas such as geology, biology, meteorology and fields not requiring experiments. China remained an agricultural country, characterized by crude agricultural methods used for a thousand years. In 1949, industry, which was dominated by labor-intensive light industry, only accounted for 12.6 percent of China's GNP and 2 percent of its workforce. Education levels were poor; 80 percent of the population were illiterate. Space programs demand technological sophistication and innovation, but neither arises in a vacuum; they depend on the mobilization of both domestic knowledge and knowledge acquired externally. The domestic constraints listed above indicate that domestic knowledge accumulation within China remained inadequate – there were too few highly qualified research personnel, with well trained technicians also scarce as support staff.

For many countries, exclusive reliance on knowledge created within the country is not wise, since knowledge from the international scientific and engineering communities is also important. External knowledge here does not merely refer to the mature technologies from other countries; rather it refers to all knowledge created or discovered by humankind. China and the Western world shut their doors to each other in the post Korean War era; the major external knowledge source for China was the Soviet Union. But after the great split in 1960, this external source was lost and China was isolated from the rest of the world's advanced science and technology. This was a huge loss for China, not only because China could not buy another Soviet rocket, but also because China could not keep up to date with the latest S&T achievements as other countries could. Thus, compared to the United States and Soviet Union, China suffered more frustrations in the R&D of space programs and its development pace was much slower than that of the two space giants.

The United States essentially closed their doors to China after the fall of the Nationalist government in 1949. Other Western states were more open to China but their export of technology and information to China was limited by their military and economic alliances with the Americans. US hostility regarding China was reinforced by Chinese support of North Vietnam during the Vietnam War and the Chinese development of nuclear weapons. The latter caused extreme anxiety in Washington, with the possibility of a preemptive strike discussed in the mid-1960s.[23] The goal was to destroy the Chinese capacity to build nuclear weapons after their October 1964 success made that threat public. The option was passed over due to political and technical issues that were deemed irresolvable.

The second era (1966–1976)

The Cultural Revolution

The principal event impacting Chinese space policy during the second era was intense domestic political strife. The second era covers the Cultural

Revolution years, when ruthless political persecutions and bloody political confrontations devastated China's economy and normal social life. According to the most conservative estimates, about 34,800 people were persecuted to death, including the president of the People's Republic of China, Liu Shaoqi.[24] Robinson is right in arguing that Mao Zedong's purpose in launching the Cultural Revolution was to protect and consolidate his absolute authority.[25] After the collapse of the Great Leap Forward, Mao found that he could no longer totally control the bureaucracy and the party. Mao complained that he could only muster a slight majority regarding critical decisions in early 1966.[26] To reassert his power, Mao launched the Cultural Revolution and in doing so purged his political rival Liu Shaoqi, along with millions of Liu's followers who were denounced as "bourgeois-liners." With the backing of the PLA led by Lin Biao, Mao eventually won the political battle by consolidating his personal power. Mao now became "the radiant sun lighting our minds."[27] "Whatever accords with Mao Zedong Thought is right, while that which does not accord with Mao Zedong Thought is wrong."[28] Mao's dominance in the political arena continued until his death in 1976. Many of the leadership cadre were purged while societal and economic institutions were totally disrupted. The PLA stood outside the turmoil to a greater extent than other government institutions. This was a major change for a self-described "socialist" or communist state since it invited the military into the inner sanctum of national power. Normally the military was kept on a short leash, as can be seen from the presence of political officers throughout the command structure.

The earlier decision to move the space program away from PLA control meant that the program was exposed to the full force of political turmoil. After a period of virtual total disruption, the space program was returned to the PLA's auspices. That realignment allowed work to begin again but the Red Guards kept the situation on edge. Budgetwise, the program was on short rations but the necessities were provided. The disruption further isolated China – to outsiders, the state appeared unstable and possibly threatening given its possession of nuclear weapons.

The space program during the Cultural Revolution years

Although China successfully launched seven satellites in total in this era, the major achievements of the Chinese space program at this stage were that the first satellite was launched in 1970, and the technology satellite was launched in 1971. China achieved the technological feat of reaching orbit but the first launch was propaganda pure and simple. The satellite broadcast a simple jingle glorifying Chairman Mao: "The east is red when the sun rises, China made a Mao Zedong. He brings happiness to people, and he is a great savior of them all." The purpose was to publicly proclaim China's success and, by extension, its heightened status within the

72 First awakenings

Table 3.2 Distribution of launches by country (1966–1976)

Country	Percentage
Australia	0.1
Canada	0.1
China	0.5
France	0.8
Germany	0.6
India	0.1
Italy	0.2
Japan	0.6
Netherlands	0.1
Russia	63.2
Spain	0.1
Sweden	0.1
UK	0.5
US	33.1

international system. Two reconnaisance satellites were launched in 1975 and 1976 respectively. The three Jishu satellites launched near the end of the Cultural Revolution years were also thought to be a pure political propoganda tool. Brian Harvey even doubts their existence, while the Chinese media no longer mentions them today.[29]

Table 3.2 indicates the proportion of successful orbital launches made from 1966 to 1976 by various countries. The proportion is calculated by dividing the launch number of each country by the launch number of all countries. The Soviet Union made the biggest number of orbital launches, followed by the United States.

If the three Jishu satellites are excluded, China lagged behind France, Germany, Japan and the UK, ranking seventh in the world in the number of successful orbital launches. Note that China and India were the only two developing countries that entered the space program club in this era. Compared to India, which made only one launch in 1975, China was the most advanced developing country in the international space club. But the development of the Chinese space program was not consistent in these years; its progress was determined by domestic politics, which disrupted the program. The space program can be divided into two distinct periods: the era of expansion before Lin Biao's mysterious death in 1971, and the succeeding period of shrinking activity.

The growth period (1966–1971)

The growth period during the second era is defined as the period during which the space program was fully supported by the highest political hierarchy, and thus the overall developing trend of the space program remained upward. This era was from 1966 until Lin Biao's death in 1971. The Cultural Revolution years were characterized by prevalent chaos. The space program was not immune from adverse effect, particularly before August 1969 when high level scientists and engineers were protected with the intervention of Zhou Enlai. The adverse effect existed. For example, two prominent scientists, Zhao Erlu and Yao Tongbin, were persecuted to death and the two political factions of the Ministry of Machinery Industry, i.e. 915 and 916, decelerated the working process. So, why were these three years still considered as the growth era of the second era? The reason was simple. The space program remained the first national R&D priority. To minimize the loss caused by political chaos, the PLA intervened in order to insure the progress of the space program by sending the Military Control Committee from April 1967. Further measures were adopted to protect the scientists and engineers working for the space program after Yao Tongbin's death. Space activities under the aegis of Mao and Lin became the only S&T sector that enjoyed high political priority in those turbulent days, while other R&D activities were despised as the work of "Stink Ninth" by the Red Guards. Compared to other S&T sectors, the loss in the space sector was much smaller, and the overall developing trend remained upward – the details of which are discussed later in this chapter.

Political coalitions between Mao and Lin also led to an amazing enhancement of the PLA's status at that time. Space R&D was returned to military control, and scientists and engineers who worked for space programs enjoyed the same prestige as army personnel. The national priority of China's missile and satellite programs was reconfirmed. Important scientists, such as Qian Xuesen, were protected from political prosecution and therefore were not humiliated and prosecuted like other individuals educated in Western countries, while China's S&T effort fell into total disarray outside the protected sectors. Training and educational programs at the universities and other centers were effectively ended by the disruptions; teachers and other trained personnel were shipped to the countryside to be politically rehabilitated through physical labor.

Space programs developed vigorously before 1971 when Lin died. Work on the DF-4 began in 1967. This missile was explicitly designed to reach Moscow and Guam. Its first successful flight was achieved in January 1970, but continuing tests did not end until 1973. Development of the DF-5, the first Chinese intercontinental ballistic missile, was initiated at the same time as that of DF-4, and in September 1971, the successful short-range test flight occurred.[30]

China's first manned space program began in 1968 under the leadership of the National Defense Science Committee. The Space Medical Institute of China, an institute that was responsible for training astronauts and doing research on manned spaceflight, was established in Beijing on April 1, 1968. The institute continued its work even when this human spaceflight effort ended without success. Nineteen a stronauts were selected from the Air Force in 1971 and their launch vehicle, the Shuguang-1, was to make its maiden flight by the end of 1973. As will be seen, this initial effort at human spaceflight was terminated without success; it was technologically premature given the larger societal disruptions.

Chinese sources indicate that work on the recoverable satellite program began in 1967. According to *Bright Daily (Guang Ming Ri Bao)*, an influential official Chinese newspaper, the purpose of the recoverable satellite was for use in manned spaceflight. In 1970, the program was listed as a project of national priority in order to accelerate the R&D processes. Concurrently, R&D on biological return missions using sounding rockets went smoothly, with a puppy successfully recovered in 1967, while work continued on achieving orbit for the first time.

Any space activities had to await successful development of a launch capability – success would also signal China's acquisition of new military capability in the form of ballistic missiles. Development of the satellite and its launcher began in November 1966. The satellite project was conducted across the Cultural Revolution years, further illustration of its national importance. The first Chinese satellite was successfully launched on April 1, 1970, from Jiuquan Launch Center, and "The east is red" was heard all over the world. National prestige enhancement was achieved with the launch of the satellite, which had an impact on Western leaders, increasing their awareness of Chinese technological progress. The launching of DFH-1 demonstrated to the outside world that China owned the "capability to build missiles for nuclear warheads." According to the *New York Times*, it is a "scientific milestone and proves the nation has developed a powerful rocket that could be used as an intercontinental ballistic missile."[31] And this "launching is expected to increase apprehension in the West over China's nuclear potential."[32] The construction of the second satellite was started in May 1970 with launch occurring in March 1971. This satellite was used for a scientific mission which lasted eight years in orbit.

Besides the space program initiatives initiated in the first era, which were continued, a more ambitious plan was made in August 1970. According to this plan, eight new launch vehicles and fourteen new satellites were to be launched over the next five years. Other space programs also started in this era, such as the Jishu (satellite) series and the FY-1 meteorological satellite. A research institute on anti-missile and ASAT (anti-satellite) technologies was established in 1969.[33]

The Mao–Lin coalition, the early space program and the Cultural Revolution

China in those years was still poor. In 1968, China's GNP per capita was only 85 US dollars.[34] But internal politics, the Mao Zedong–Lin Biao political coalition, explains the expansive trend over the period May 1966 to September 1970. To fully understand the motivations for the Chinese space program during this time, one must examine the Mao–Lin political coalition's goals.

General Lin Biao, after lengthy service in the PLA, started his public role as politician in 1959 when he became the Defense Minister. Talented in the art of war, Lin was also proficient in political maneuvering. While other parts of China under Liu Shaoqi's leadership were rejecting Mao's radical ideas after the Great Leap Forward failure, Lin on the contrary was vigorously indoctrinating the PLA with Mao Zedong thought. His political affiliation was clearly expressed in his first public speech (October 1, 1959), with the title "Hold High the Red Flag of the Party's General Line and Mao Zedong's Military Thinking, March Forward in Mighty Strides."[35] Lin bet his entire political stake on Mao Zedong and mobilized all the resources he controlled to enhance Mao's position, including deifying Mao. Speeches included "The Victory of the Chinese People's Revolutionary War Is the Victory of the Thought of Mao Zedong,"[36] "Chairman Mao has Elevated Marxism-Leninism to a Completely New Stage with Great Talent."[37] These public pronouncements indicate that Lin had made a huge political investment in the declining power of Mao at that time. In October 1961, the PLA issued the first edition of the "Little Red Book" which was later promulgated to the whole country. The "Little Red Book" was a compilation of Mao's slogans and excerpts from his speeches, materials aimed at fomenting and encouraging greater militancy among the Chinese people, especially the young.

Because of Lin's contribution to his political power, Mao in February 1964 launched the movement called "Learn PLA."[38] Although Mao's power was dwindling among the inner circle of top civilian political leaders, his personal prestige as a revolutionary was still high and strongly rooted in lower level officers and poorly informed masses. Thus the political coalition between Mao and Lin proved a win–win game for both sides. Lin, with the image of being both "red" and "professional," became better accepted by the masses due to Mao's "Learn PLA" movement, while Mao benefited from Lin's further promulgation of Mao Zedong thought within the army, the ultimate power source in an authoritarian regime.

Lin played a crucial role in Mao's battle against Liu during the Cultural Revolution.

> As the struggle between the Liuist and Maoist forces drew to a head in the late spring and early summer of 1968, Lin managed to assure control of the Peking military garrison and to disarm or neutralize units of the public security force in the city.[39]

Again, "military intervention was a necessary and the most effective instrument for the dominant leaders to solve a policy crisis."[40] Shortly after, at the Eleventh Plenum in 1966, Lin replaced Liu as Mao's designated successor. The PLA's political intervention resulted in its predominant role in Chinese politics across the 1960s. Not only did Lin and his most intimate followers gain major positions in the central government, but the PLA as a whole emerged as a dominant political force and it, in fact, replaced the CCP in ruling the country.[41] The PLA's position in Chinese politics reached its apex in April 1969 when Lin, as Mao's successor, was written into the Constitution, while PLA representation on the Central Committee rose from 19 to 45 percent.[42] The quick military victory over India in the 1962 border conflict, the 1964 explosion of the atomic bomb, the Vietnam War and the open Sino–Soviet split, all provided the PLA, with Lin as its commander, with additional political weight.

R&D on the space program was put under direct military control. This turned out to be the golden period for technological breakthroughs in early Chinese space program development. The R&D decisions regarding DF-2, DF-3, DF-4, DF-5, the first satellite (DFH-1) and the second satellite (Shijian) were all made in this era and they were all successfully launched. Likewise, the R&D decisions regarding development of a recoverable satellite and the Jishu satellite also occurred at this time.

Rapid space program development reflected Lin's political power in obtaining social resources for the military sector and his desire to build and modernize the PLA. He clearly knew that the PLA was the best card by which to win the game of political competition with other political elites. Being the most talented Chinese military commander of the twentieth century, Lin was proficient in military affairs and certainly knew the importance of advanced weapons. Verbally he supported whatever Mao said, including the concept of a People's War which emphasized that man is more important than weapons, but, in fact, Lin avidly supported the development of modern weapons. Mao rhetorically proclaimed that men are superior to weapons; that was because China did not otherwise possess enough weapons to deter the West or the Soviets. This verbal approach also played on American apprehensions after the Korean War about becoming bogged down in an Asian land war, apprehensions that the Vietnam experience was reinforcing. What else can a paramount leader do but verbally encourage his people publicly while pursuing such weapons covertly? From Mao's perspective, he was moving to consolidate his power with the PLA's support and thus he strongly supported space programs under the PLA's leadership. Space programs almost by definition have futuristic implications, especially military, since the field is totally open to whatever fantasy one wishes. Thus the rapid development of the space program in the 1960s and the early 1970s arises from the fact that the space program became a tool for a political faction to consolidate power. Lin's contribution to the rapid development of the Chinese space industry is obvious, but his contribution is seldom mentioned, because after his death in 1971, Lin was considered the most traitorous person in Mao's faction. He was also bitterly opposed by Deng, the second-generation leader. To Deng's faction, Lin was a criminal for assisting Mao in deposing and prosecuting Liu and Deng's faction. Even scientists under his patronage were either too frightened or too a shamed to objectively assess Lin's contribution to the Chinese space program. It was considered too politically hazardous.

National space programs, especially in their early days, are often trophies for the dominant political faction – that was clear both in the United States and in the Soviet Union. The John Kennedy administration exploited the imagery of space exploration as part of their political message, while Nikita Khrushchev, the Soviet leader, clearly employed the Soviet space "firsts" to bolster Soviet prestige internationally but also domestically. But, in both instances, the identification was never so strong that the space programs suffered after a political transition. Clearly, the expansive days of the early space age ended in both countries, especially after the July 1969 lunar landings, but the programs pushed on expanding their range of activities. In the Chinese case, the space program was so personally identified with a fallen political figure, Lin Biao, that it entered a period of decline as opponents of Lin and Mao assumed control. The space program did not end but rather entered a period of stagnation compared to the promises and plans of the Lin period.

The period of relative decline – the running down of the Cultural Revolution

Space programs declined, at least comparatively, after Lin Biao died on September 13, 1971. Compared to the growth period, space activity in this period was reduced. The only major achievement between 1971 and 1976 was the launch of the recoverable satellite in 1975, four years later than the program had planned. Only two new space projects were started during this time – a communication satellite and the Shijian-2 S&T satellites (both were initiated in 1975). The primary concern of space policy making in this period was not for the general good but mainly for bolstering Mao's personal political position, which can be seen in the decline after Lin's fall and Shanghai's ascending position in space R&D. Shanghai had become an important power base for Mao, which meant resources and programs flowed that way even though technologically Beijing had the stronger technological base.

Conflict between Mao and Lin arose in March 1970, when Lin proposed to restore the position of head of state, which Lin saw as his rightful position. At the end of August 1971, this conflict became heated when Chen Boda, Lin's core supporter, was publicly criticized. To flee from Mao's retribution, Lin, according to the Chinese official explanations, planned to fly to the Soviet Union after an aborted coup, but his plane mysteriously crashed over Mongolia in September 1971. Then, the witch hunt against Lin's faction began. In the Ministry of Machinery Industry, the core of the Chinese space program, even the lowest ranking engineers were purged due to the great support they had received from Lin. The minister of the Seventh Ministry, Wang Bingzhang, was put into prison for ten years because of his close relationship with Lin. (Wang was the Associate Director of the Fifth Academy before 1962; after 1962 he was in charge of the Chinese program for nine years as the administrative head and made a significant contribution to China's space program.) Qian had to make self-criticism for a comment supporting Lin. Even the vibration bed that was designed to train astronauts was charged with the vicious purpose of curing Lin's insomnia.[43] Further discussion of China's political mechanisms is helpful in understanding this great change of personnel.

Historically the pyramid political structure has existed in China for 5,000 years, with the paramount leader at the top.

> All incumbents in China are appointed by his superior instead of winning his office through an open competition. There are few rules and regulations for cadre promotion. What really counts in a cadre appointment is the appointee's *guanxi* (personal relationship) to his superior; ability and virtue are of secondary importance.[44]

The whole society, therefore, is like a huge personal relationship network woven from the center and spread to the periphery. Thus, it is easy to understand that if a top official falls, people under his protection will fall like dominos.

It is not surprising that the space program under the direct charge of the Seventh Ministry was either suspended or abolished. The ambitious plans made in August 1970, to develop eight new launch vehicles and fourteen new satellites in five years, disappeared. The first successful short-range test of the DF-5 was made on September 10, 1971, two days before Lin's death, but the long-range test was suspended until 1977, one year after Mao's death. R&D on the DF-6 launcher and most notably the Shuguang-1 manned space program was simply cancelled, while development of the communication satellite was postponed. All in all, space R&D programs withered after Lin's death.

Thus, the first Chinese manned space program ended in a whimper with no actual flights occurring. Politically speaking, ending it at this point was not as great a setback as would have been the case if cancellation had occurred after flight mishaps. The Soviet and American programs were able to persevere because they had a history of success; failure was not as devastating. China's status in the international space realm was such that the cancellation did not impact it. Other than the Americans and the Soviets, no other state had even broached the idea of an independent human spaceflight effort, although some were linking up to the US shuttle program which got its go-ahead in January 1972.

The second proof that space programs in this period focused on Mao's personal power concerns was Shanghai's ascending role in space R&D, symbolized by the construction of the Jishu satellite series and the Feng Bao launcher. From 1966 onward, space R&D tasks were assigned to the Shanghai team, although the Beijing team had major competitive advantages regarding both capital equipment and labor resources. According to Harvey,[45] Shanghai had no test equipment while other critical components were also poorer in quality. But from the beginning of the Cultural Revolution, in support of his own political goals, Mao committed significant resources to shift some space programs to Shanghai. Harvey believes that this was because Shanghai was Mao's political base and Shanghai was politically safe to Mao.[46] Although Mao might have thought that he should not put all his eggs into one basket and that the space program should be under different groups to maximize the probability of his ultimate control, the more direct and reasonable rationale was still resource control competition among political elites. Note that Lin Biao was still Mao's "comrade in arms" in those years and Beijing was tightly controlled by Mao and Lin. Shanghai's ascending role in space R&D cannot be explained simply from the perspective of being politically safe. Remember that military control is crucial for political elites' survival in authoritarian countries, and space technology represents the most advanced military technology. Thus space technology becomes an important resource that political elites compete and fight for. This was particularly true during the early years of the Chinese space program, which stressed its direct military applications and prestige enhancement. Note that the dominant players in the Chinese political arena changed dramatically after 1966, when the Cultural Revolution started. Zhang Chunqiao and Wang Hongwen became the new salient political upstarts strongly supported by Mao, and they were both from Shanghai. The Shanghai faction thus got a big political leverage and won their share of space technology control. Shanghai's position was further enhanced after Lin's death in September 1971, because the Beijing team was estranged from Mao due to Lin's avid support of their efforts. Poor capital investment and insufficient S&T investment resulted in a series of failures in the short-lived Jishu satellite series. No public information was ever released regarding the possible usage of this satellite series; it was possibly of no other use than radical political propaganda, since few contemporary Chinese official review articles mentioned it, while all other space projects are listed as achievements. R&D on the Jishu series was possibly repetition of what the Beijing team had done years before, but the construction of the first Jishu series took the Shanghai team five years and all the Jishu satellites functioned poorly. The first Jishu satellite was launched after a series of failures in July 1975, and it burnt up after fifty days. The second satellite was launched in December 1975 and it decayed after only forty-two days. The third one was launched in August 1976; it survived in orbit for 817 days. All three satellites were given great political significance by the Gang of Four, the radical Maoists headed by Mao's wife (Jiang Qing), in their political campaigns against Lin Biao and later Deng Xiaoping. Even the slightest technological information, such as orbit parameters, was never provided, and the West never picked up their signals.[47] This was similar to the episode when North Korea claimed that it had orbited a satellite that no one else observed or heard from orbit.

For the space program, this was the tragedy of the Cultural Revolution years; the whole society revolved around the vicissitudes of a certain person or a small group's desires rather than the work's quality or usefulness. All space activities had to yield to the political struggle; consequently huge social resources were wasted in a society where such resources were always scarce. Chinese space programs were hugely adversely affected in the latter period of the Cultural Revolution. If the Mao–Lin coalition had not collapsed or if the disruptions had been limited in time and scope, the Chinese manned space program might have been achieved decades earlier (the first Chinese manned spaceflight was accomplished in October 2004). Due to political difficulties, the Chinese space program lost momentum while the nation's elites sorted out their relationships. Much awaited the departure of Mao from the political scene in 1976.

Economic and technological constraints

Continuous political turmoil destroyed the basic order of society and economic development stagnated or even decreased during the Cultural Revolution years, especially in the early years until 1975. In 1967 and 1968, the industrial and agricultural index decreased by 9.6 and 4.2 percent respectively over the previous year. The economic gap between China and the outside world grew larger. The overall GDP per capita over the ten years was $789.7, about one-twentieth of that of the United States. Note that in the first era, China's GDP per capita was one-eighth of that of the United States. Clearly, the Chinese macro-economy deteriorated during these years while the United States continued to grow. The grand plan of the growth years was terminated due to political factors. If those plans had been carried out, China might have followed the Soviet road and depleted other societal resources to support its ambitious space programs, which in the Soviet Union eventually led to the regime collapse.

Besides the first satellite that was launched in 1970, the most prominent achievement of this era was development of the recoverable satellite. Recovering a satellite from earth orbit was more challenging than simply launching one without recovering it, due to the engineering difficulties such as protective heat shield design, retrorocket, precise control and ground tracking system. For the Chinese scientists and engineers at that time, the recoverable satellite was both an incremental and a breakthrough innovation. The R&D of recovery satellites was started in 1967, but the first attempt to launch was not made until November 1974, and the debut launch ended up as a failure. Finally, after about nine years' hard work, the recoverable satellite was successfully launched in 1975. This nine years' endeavor itself indicates the large technological constraints impacting China. Note that the United States successfully launched its first film-return type of reconnaissance satellite only one year after its first satellite launching, the Corona program. The Americans endured thirteen failures before achieving success, but they had the resources to expend in such a pursuit. The Soviet Union took a longer time, but it still took only about five years. During the nine years of recoverable satellite R&D, a series of engineering problems bedeviled Chinese scientists and engineers. For example, retrorocket manufacturing required major testing before the systems operated properly. In a society stretched for resources, such testing came at a high price, tying up precious technical personnel longer than projected.

China could by no means make the same incremental innovations in space technology as quickly as the United States and the Soviet Union. The resource differences were too substantial. Recall that after the first satellite in 1957, new types of orbital launches appeared every year for both the United States and the Soviet Union. By 1965, the United States had launched 11 communication satellites, 8 ELINT satellites, 20 meteorology satellites, 18 navigation satellites and 106 reconnaissance satellites, plus manned spaceflight, lunar probes and other deep space missions. Similar patterns existed for the Soviet Union, although the early Soviet space program emphasized national prestige programs rather than applied satellite series. Scientific and technological innovation in China remained slow and China lacked both incremental and breakthrough innovations for a time.

As discussed before, scientific and technological innovation depends on both domestic knowledge and external knowledge, and the flow of both was interrupted by the radical policy adopted in the Cultural Revolution years. Internally, the Cultural Revolution destroyed the national research and education system. Researchers in most institutions stopped working during these years; instead they were sent to the rural areas to learn political virtue from peasants, since, according to the doctrines of the Cultural Revolution, workers and peasants were the main force of revolution while intellectuals did not hold firm revolutionary beliefs. Many scientists and engineers were publicly denounced and some even persecuted to death. For example, Zhao Jiuzhang, a German Ph.D. and a major contributor to sounding rocket and satellite research, committed suicide at his home due to political persecution. Yao Tongbin was simply killed by Red Guards.

"China lost its opportunity for rapid technological development by destroying rather than building on its past accomplishments."[48] Teaching was halted from 1966 to 1969 so that students and teachers could devote their energies to the Cultural Revolution. The education policy adopted afterwards downgraded academic standards. The length of education at each level was cut by at least a year and college admission was based on students' political qualifications rather than their academic performance. Relations with the Soviet Union remained as bad as before and the relationship with Western countries was ameliorated but not good enough to commence science and technological exchanges. Moreover, radical slogans such as "we would accept grasses of socialism rather than seedlings of capitalism" severely misguided people into rejecting the very limited external knowledge available. The Cultural Revolution for ten years created a huge knowledge faultline for Chinese generations born in the late 1940s and early 1950s. This faultline had a huge adverse impact on later space program development since technicians and scientists were not available in the numbers and of the quality required to advance in the space realm.

Conclusion

Space programs during these first two eras were highly centralized and basically military-oriented. Missile programs developed quickly in the 1960s and, by October 1971, China could launch an ICBM and thus insure China some capacity for strategic nuclear deterrence. The twenty years from 1956 to 1976 witnessed full-scale internal power struggles and potentially dangerous international confrontations. Internationally, China had to confront the superpowers, first the United States and later the Soviet Union. Military-oriented space programs, especially missiles, met the urgent need to counter the superpowers' "nuclear blackmail," which China continuously confronted in the 1950s. Therefore, the first motivation that drove the vigorous development of the Chinese space program in these two eras came from external threats. Domestically, the first two eras were full of ruthless political persecutions and bloody political confrontations. At one stage, Mao and Lin formed a political coalition of "charisma and iron and blood." With the help of Lin's military forces, Mao consolidated his undisputed rule in 1968; meanwhile, Mao's charismatic support enabled the PLA headed by Lin to become the dominant political force within China in the late 1960s. Missile programs, an indivisible part of the PLA, brought Lin more political capital; meanwhile Mao's supreme leadership also relied on the army. Personal political need made Mao woo the space program. Thus, the eleven years from 1960 to 1971 marked the golden period of space development. After Lin's political fiasco and death in 1971, Mao purged Lin's faction and the space programs he patronized could not be saved; consequently the Chinese space program declined after Lin's death. Thus, the second motivation for Chinese space program development was to serve as a political tool for consolidating personal political power. In addition, space programs were used as a propaganda tool, which further stimulated the authoritarian leader, Mao Zedong, to support these multifunctional programs, particularly when the economy was in dire straits and the Chinese people needed a psychological boost. Overall, space program development in this period arose out of the urgent needs of national defense, national prestige enhancement, and internal political utility.

Chapter 4

Accelerating the rise of China's space program

Introduction to the third era (1978–1986)

The third era runs from late 1976 to the beginning of 1986. The major theme within the space program in this period switched to advancing Chinese economic development with the primary focus on the civilian side of the dual-use satellites and space program commercialization. This switch occurred as the result of fundamental changes occurring in China's elite politics. Mao Zedong's doctrine that class struggle must be the political core value of the country was gradually replaced by Deng's more pragmatic philosophy that economic development should become China's priority. Deng's pragmatic approach became dominant within China particularly after 1978 when he became undisputed leader. Political resistance to Deng's policies persisted for years but opponents have been unable to stop rapid economic change from occurring.

Development of the space sector clearly reflected this broader political change. For example, China launched three major series of satellites: DFH (Dong Fang Hong) communication satellites, FWS (Fan Hui Shi) recoverable satellites, and SJ (Shijian) science satellites. Although these satellites were dual-use in nature, the Chinese government publicly emphasized only the civilian-side use, for example using telecommunications for improving education.

In addition, China seriously opened its doors to the Western world for the first time since 1949. China was selectively breaking out of its isolation. The purpose was to acquire technologies and training – there was no official interest in Western political values or views although in the end those also infiltrated the society. As one of the few high-tech sectors that was internationally competitive, the space sector was thrown open to the international market as a means of earning foreign currency, which was of great importance if China was to purchase other technologies and equipment. In 1985, as more Long March launch vehicles were developed, China started providing commercial satellite launching services to international customers. The CZ-3, a geosynchronous equatorial orbit launcher, was first launched in 1984, while work on the other two launchers, the CZ-4 and CZ-2E, was also initiated in this era. After the serious disruptions of the previous period, China's space program had to first restore its technological base, especially with regard to trained personnel.

Restarting the effort meant replacing the older generation with new trained personnel – these individuals had to be indigenously trained since China during its political convulsions had become isolated from much of the global scientific and engineering community. They had to acquire both high quality training and relevant experience – the latter could come only through doing. The relative political stability created by Deng's accession gave the opportunity but the political pressures remained real. All resources needed to be used efficiently and effectively since China's societal needs across the board were so great. China drew prestige from its space activities but one could not live on accolades; instead, tangible results had to accrue.

The third era saw China's space program enter what might be termed a period of relative normalcy. That is, the Chinese space program began moving into the space applications development mode. Political prestige still remained important as a symbol of China's rising status in the world but the program now had to justify itself beyond that narrow range of motivations. In this, China tracked the American path of space commercialization in that practical applications became important, but with one important difference. In China, the government retained control and the means employed were governmental – no private sector existed. The Soviets moved more slowly in that direction. Soviet slowness did not reflect backwardness but rather the contrary. The Soviets considered themselves and were considered by others to be an advanced state. Therefore, the Soviets lacked the motivation, both ideologically (being Marxist) and economically, to make such a dramatic change. For the leading communist or socialist state, commercialization was an admission of failure. Only when the system was in severe economic disarray by the 1980s did the necessity for change become overwhelming. The Soviet Union fell within the decade, too little, too late.

For China, once commercialization was initiated, it was aimed externally rather than internally, but the implications of such a change were clear despite the political rhetoric that continued affirming the old Marxist certainties. Change was incremental rather than rapid – the space program exhibited that same pattern. This comparative stability allowed systematic development of new technologies – the new pattern within the program was to proceed forward incrementally rather than to make mad dashes to nowhere. This Chinese characteristic comes through clearest in the human spaceflight program with its step-by-step approach to the problems, including steps backwards. The third era saw China push forward slowly because the leadership kept its expectations under control; political discipline became the hallmark of the program.

Satellite programs

As indicated, three series of satellites became the focus during this period. These three, plus a fourth, which drew less attention, were: the DFH (Dong Fang Hong) communication satellites, the FWS (Fan Hui Shi) recoverable satellites, the SJ (Shijian) science satellites, and the FY (Feng Yun) meteorological satellites. Development of communication satellites was a high priority, since China's vast territory and diverse landscapes make communication among different areas very difficult. Two-thirds of China's territory is mountains and plateaus, including the world's highest plateau (the Tibetan Plateau) with an average elevation of 14,800 ft, which dramatically increases the cost of cable communication. China is similar to Canada with its distant and scattered northern population – a fact which led Canada to pursue comsats as a means of linking such populations to the core regions.[1] The first successful experimental communication satellite was launched in April 1984, followed by the first operational communication satellite (DFH-2 series) launched in February 1986. The early communication satellites were technologically simple with only two transponders, but the antennas provided beams that could cover the whole of China. They provided effective communications to border regions such as Xinjiang, Xizang and some islands off southern China. The DFH-2 remained operational until July 1989. The communication satellites of this era were equipped to provide television coverage and transmit telephone and fax data.

Communication satellites serving China already existed but those were controlled by the Western states through Intelsat. Their technologies, especially Intelsat, were much more advanced than their Chinese counterparts but the DFH program provided what was essential – communications independent from outside control. China geographically is a vast country which contains significant physical barriers to communications – independent comsats allowed Chinese authorities to access those regions without interference. For a state whose enemies or would-be enemies were powerful, this was considered a priceless asset.

China's policies regarding comsats resembled those of the United States and tracked a pattern of government-controlled and -operated satellites, joined later by commercial comsats leased by the Chinese or operated by international corporations or consortiums. The United States for example has flown a series of comsats solely controlled by the military for critical communications, with commercial comsats handling less critical messages. Once China demonstrated the capacity to successfully build and operate comsats, it moved to a higher level of space applications.

Work on the FY-1 and FY-2 meteorological satellites commenced in 1978 and 1980 respectively. Besides weather forecasting, these two satellites were also designed to monitor rainfall and flooding, crops, forests, and collect evidence of pollution. Meteorological satellites are important for a state geographically as large as China – flooding and drought are two recurring disasters in Chinese history and still are a direct threat to the countryside, where 80 percent of the population rely on agriculture for their living. Such satellites allow the government to respond in some fashion to the disasters, helping mitigate their effects by early warning. Militarily, weather information is important for conducting operations.

The recoverable satellites project which began earlier during Mao's tenure continued in this era, but their principal mission was primarily civilian in nature. By October 1985, four FWS 0 satellites had been launched and recovered successfully. The FWS 0 series was used for land surveys. China also developed and launched a science satellite, the Shijian 2, in 1981. This spacecraft carried a cosmic ray detector, X-ray detector and magnetometer in pursuit of useful information on the Earth's radiation belts in furthering space exploration. This emphasis on radiation belts is typical of new space programs since that hazard proves extremely disruptive when they pursue more advanced applications.

China had clearly moved beyond the simplicity of "the east is red" message of the early 1970s. Ironically, China now started the process of building a space program that other states would have to notice and respond to. How those states would respond was not entirely clear. Isolation breeds misinformation which raises concerns on both sides, China and the others.

Development of launch technologies

The launch vehicles developed in this era included the Long March CZ-3, CZ-4 and CZ-2E (see Appendix A for data). As mentioned, the CZ-3 was designed to launch communication satellites to geosynchronous orbit (GEO). This three-stage launcher, with a launching capability of 1,500 kg (GEO), was first flown in 1984 and became the first public symbol of the improved high-tech product being produced in that era. One distinct characteristic of this launcher was that its third stage used a liquid oxygen/ liquid hydrogen engine, which enhanced engine performance and made the combustion products non-toxic. This launcher allowed China in 1985 to enter the international commercial launch service market.[2] By June 2000, the launch success rate of CZ-3 was 85 percent. Development of the CZ-4 started in 1979 with its mission profile being to launch the FY meteorological satellite series to polar sun-synchronous orbit.

During this era, China also demonstrated for the first time its multipayload launching technology. China was never first but its efforts distinguished it from other underdeveloped states. On September 20, 1981, China launched three science satellites using one launcher and became the third country after the United States and Soviet Union to possess this technology. The Europeans followed suit later with their Ariane rocket technologies.

Their launcher program had begun earlier in the 1960s and early 1970s and had early on suffered from political uncertainties, not as drastic as in China, but disabling at times.[3]

Opening up to the world

Compared to the earlier two eras, the distinctive feature of Chinese space programs during this era was its opening up to the outside world in ways considered unacceptable previously. In contrast to Mao's closed-door worldview for reasons of ideological purity, Deng Xiaoping viewed this opening of China to the outside world as the necessary shortcut to updating China's science and technology endeavors.[4] International R&D exchanges in the space sector developed quickly with strong government encouragement. In 1980, China joined the International Astronomical Federation, International Telecommunications Union, and the Committee on the Peaceful Uses of Outer Space. In 1985, the first international conferences on space applications and policy were being held in Beijing.

In that same year, China signed a contract with Brazil to jointly develop two remote sensing satellites, the Ziyan (resources) series (also known as China–Brazil Earth Resources Satellites (CBERS)), which were designed to collect images and other data that would be used for planning, surveying, agriculture and disaster monitoring. After the launch of the DFH-2 in 1986, the State Council decided to work with foreign companies to develop C-band communication satellites. These decisions demonstrated China's great interest in becoming integrated into international society, especially in the high-tech realm, in which each side received benefits from exchanging information, although China clearly benefited most initially.

China pursued this new policy for two reasons. First, Chinese space efforts had reached a level of sophistication where cooperation was obviously necessary if their rapid progress was to be sustained. The Chinese could now identify those areas where their deficits were greatest and where others might be willing to provide the requisite knowledge or technology. The loosening of Cold War security restrictions after the collapse of the Soviet Union in 1991 made such technologies more available but even in the 1980s such exchanges became possible. The United States and the Soviet Union were no longer the dominant players in many technological areas; other states, especially in Europe, had developed equivalent or useful technologies of their own and they were available. At the same time, growing global commercialization of space applications further loosened the export restrictions placed on China.

Second, the Chinese now possessed the self-confidence to, if pressed, develop the technology by themselves. Therefore, assistance would be perceived by the Chinese as an exchange between comparative equals rather than the superior–inferior relationship of its status during European semi-colonialism. For a state that suffered much under that colonial status, being the inferior was now politically unacceptable. The other side of the coin was that as others became more cognizant of Chinese capabilities, their nascent efforts at conducting commercial activity would gain greater credibility. This can be seen in the George H.W. Bush administration's 1990 decision to allow American comsats to fly on Chinese launchers. At that time, US comsats were the world standard for that application; the Chinese launch industry had to raise its game in order to compete successfully for any launch contracts. The progress made during the third era was the basis for such decisions. The quality was comparable so the competition would be on price.

Overview of the third era

The third era was very different from the first two. In order to acquire a real understanding of the underpinnings of Chinese space policy after 1976, it is first necessary to undertake an overall review of this era. Remember that in China the ultimate decision-making power regarding the space sector has been controlled by a few leaders at the top of the government hierarchy. Since these political veto players usually did not understand the technical matters related to space technology, they relied on the information provided by their bureaucratic agents to make decisions based on their calculation of what was necessary or even possible.

Government space programs are never insulated from the effects of political choices, and often individuals make difficult decisions based on only partial understanding of the choices. The quality of the advice and information received from those actually operating the program ultimately determines the quality of the final decision. Stakeholders usually exaggerate their successes and minimize their problems – an expected reality – in communicating to their superiors. The key variable becomes the leaders' understanding of the larger world context and how this new information fits into their worldview, plus their openness to information contradicting their views.

During the first two eras, the latter was not possible. Uninformed decisions or ones marred by strongly held misconceptions can severely damage a space program, as occurred during Mao's tenure. Thus, the top leader's worldview and political calculations determine the overall policy outcome. In the first and second eras, the direction of China's space policy was ultimately decided by Mao Zedong, while in the third era the major decisions rested with Deng Xiaoping, who served as effectively undisputed leader from 1978 to the mid-1990s. The Hua Guofeng period from Mao's death in 1976 to the Third Plenum in 1978 was the transitional period prior to Deng's reforms.

Deng Xiaoping was born in 1904, eleven years later than Mao who was born in 1893. His early experiences in France made him more informed regarding the Western world. His well known saying is that "it does not matter whether a cat is white or black. As long as it can catch mice, it is a good cat," which has often been interpreted as it does not matter whether we have socialism or capitalism. As long as it provides people with what they need, it is acceptable. Deng was not only pragmatic in his thought, but also demonstrated his talents for fostering economic development. China's focus under his leadership would be economic development and that effort incorporated the national space program. Previously, under Mao, the space program was disconnected from economic development; the focus was prestige and power projection.

For Deng, focusing on economic development was a win–win situation. China's economy was virtually on the brink of national bankruptcy when Mao died in September 1976. The Cultural Revolution shook people's confidence in the government and lessened their support for the Party. "The post-Mao leadership realized that the legitimacy of the regime would hence forward have to depend on its ability to deliver the goods to improve people's living standards through the modernization of agriculture, industry, national defense, and science and technology."[5] The probability of economic development created by economic reform was almost certain. The people did not have to have a deep knowledge of economics; the fact that a private-oriented economy would work much better could be easily seen from comparing several pairs of countries, say Taiwan and the mainland, South Korea and North Korea, or West Germany and East Germany prior to reunification. In Deng's calculations, successful economic development would enhance his power against his political rivals (mainly the radical or traditional wing of the Communist Party), and therefore allow him to hold power. Overcoming the radicals, however, took time and proceeded in stages. Different from Mao's era, this time the interests of an ambitious politician coincided with the interests of the larger society. The national priority was shifted from Mao's class struggle to economic construction.

Economic construction became the major theme of this era and achieving the Four Modernizations became China's new national mantra. The Four Modernizations refer to modernization occurring in agriculture, industry, science and technology, and defense. They had first been enunciated by Zhou Enlai in January 1975 but were taken up by Deng as his political program in opposition to the Maoists. By the end of his life, Mao on principle opposed modernization with its emphasis on material progress. He died in September 1976 and his political heirs (the Gang of Four) were unable to prevent Deng's reemergence as leader, although their resistance and that of others continued across this period of Chinese space policy. By August 1977, the Four Modernizations program was accepted as the blueprint for China's future.

The program exerted a far-reaching influence on Chinese society. First, the Four Modernizations legitimized privatization reforms in both agriculture an economic process.[16] Many believe in microgravity materials processing as a viable option but are stymied by the high costs to orbit. Those costs keep such products noncompetitive in the marketplace.

As part of the dramatic shift in government policy, China began to explore the possibilities of commercialization of its hard-won technologies. This pathbreaking move by a Chinese government still publicly avowing Marxism was driven by their desire to economically reform the country. China's space technologies could become a source of hard currency which would further their ability to purchase Western technologies. Those purchases would allow further upgrading of Chinese technological capabilities. Ironically, in this same time frame, the Soviet Union under Mikhail Gorbachev began commercializing their launch technologies. The same logic impressed the Soviets as it had the Chinese; becoming economically competitive required resources and access to Western technologies, especially computers.[17]

In that vein, from 1980 onward, the Great Wall Industry Corporation began the effort to commercialize Chinese space technologies. The FWS payload services and commercial launch services became the twin windows of opportunity for Chinese high-tech exports. Other Chinese technologies lagged behind world standards – making them noncompetitive on the market. According to Wang Liheng, the Vice Administrator of the China National Space Administration at the time, the goal of the Chinese space program then became accelerating the commercialization of its space technologies.[18] Both commercial launch services and payload services employing recoverable satellites could potentially provide China with considerable foreign exchange. Each launching can provide China with millions of dollars. For example, China earned $30 million for its first commercial launching and $300,000 for its second one, which was significant for a country with a GDP per capita of less than $300.

Space programs for military purposes and national prestige enhancement were not the major theme of this era although they were not totally absent. In his quest for national economic development, Deng generally weakened the PLA's political influence during these years. Deng made it clear that China must first develop its national economy and raise its living standards before military modernization could occur.

> The four modernizations should be achieved in order of priority. Only when we have a good economic foundation will it be possible for us to modernize the army's equipment. . . . What we have to do now is to put all our efforts into developing the economy. Economic development is the most important, and everything else must be subordinate to it.[19]

Deng convinced the Chinese leadership that a major war against China would not break out once the Soviet threat ended in 1983. Therefore, China would experience a peaceful international environment while it pursued economic development. According to Deng's analysis, "the U.S.S.R and the United States were locked in a global military stalemate, a nuclear war between China and the Soviet Union is extremely unlikely."[20] In fact, since the 1972 visit to China by President Richard Nixon, the United States and China had cooperated politically (albeit not formally) in opposition to the Soviets. China had become a balancer of sorts. Once the Cultural Revolution ended with Mao's death in 1976, the two states, China and the United States, had further "normalized" their relationship, reducing the general tension levels except when Taiwan entered the equation.

The important formal leaders during this era, Zhao Ziyang and Hu Yaobang, were both keen supporters of the light-industrial sector and political liberalization reform. Military expenditures from 1983 to 1986 decreased continuously each year. In the 1960s and 1970s the average military budget was about 19 percent of the total; by 1985 it had declined to 10.3 percent. Military modernization remained modest and absolute priority was given to fostering civilian economic sectors. In 1985, one million soldiers were demobilized, and selected military facilities began to be transferred to civilian uses. "In terms of resource allocation, the modernization of national defense receives the most unfavorable treatment among the Four Modernizations."[21]

Yanping Chen indicated that the Chinese space program was put on the back burner in this era.[22] Although the space program was not given a prominent position during this era, solid and necessary foundations were laid for the vigorous development of the Chinese space program in the fourth era. First, in this era, China laid a solid economic foundation. The Chinese economy moved out of the swamp of stagnation and developed quickly due to ongoing economic reforms. The GDP per capita by 1985 was almost twice that of 1976, when Mao died. The economic gap between China and other countries was significantly reduced. In 1976, the GDP per capita of China was about one-nineteenth of that of the United States, but in 1986 the gap had shrunk to one-thirteenth. More importantly, the economic reforms started in this era laid the cornerstone for the Chinese economic miracle sustained until the present. Economic constraints were ameliorated and the economic accumulations occurring in this era laid the necessary foundation for China's ambitious space programs as the twenty-first century approached.

Second, this era created the necessary manpower foundation. During the Cultural Revolution years, an entire generation lost the opportunity to be educated in universities or other professional schools. Labor and knowledge constraints consequently became a major threat to future space program development in the 1980s. The education policies established during the Cultural Revolution years were abolished when Deng came to power. For example, the College Entrance Examination process was resumed with school admission again based on students' intellectual potential and performance. Moreover, a standard educational framework was set for high schools and middle schools across the country, which made China one of the leaders in math ematics education. This proficiency provided the foundation for advanced technical and scientific training. Meanwhile, China started sending students to study abroad again. Over 35,000 students were sent to North America and Western Europe between 1979 and 1986. Those who returned from abroad together with those who graduated after the 1950s within the country became the backbone of the Chinese scientific community. With the guidance of senior scientists, who were usually trained abroad prior to 1949, these individuals made great contributions to the Chinese space programs in this era and up to the present.

Third, this period also laid an institutional foundation in space program management. Internally the old chief designer system was restored and the Space Ministry established. The chief designer concept provided both leadership and focus to China's space efforts – overcoming the disruptions of the past. The increasing openness of the space program helped minimize the one potential defect of this office which was that a single individual could divert resources toward dead ends. The chief designer was a derivative of Soviet science and technology policy, which worked when the individual was a strong leader with vision but floundered when lesser individuals occupied the office. Secrecy bred inefficiency and wrong decisions that often went unchallenged for too long, which ultimately damaged the Soviet program.[23]

Externally, Chinese S&T cooperation with other states resumed but on a larger scale – larger both in terms of resources and number of cooperative relationships with diverse states. The cooperation with Brazil on the Ziyuan resource satellites (the CBERS program) and with foreign companies on the C-band communication satellites were prominent examples. China also entered the international commercial launch market, which was important for forcing the Chinese to raise their performance level before foreign investors would entrust their costly comsats for launch on Long March vehicles. With these exchanges with the outside world, China began updating with the new technology and the new sources of knowledge, all of which helped ameliorate the earlier technological constraints that had plagued China.

Economic and technological constraints

The space program as a whole over these ten years was not as active as over the later years. The Chinese economy largely continued as a planned economy. The economic reforms were concentrated in the coastal regions, especially the enterprise zones. That meant political support and economic constraints were negatively correlated, i.e. strong political commitment meant fewer constraints. It makes no sense to explore economic constraints without analyzing the level of political support achieved. Support from the national leadership for the space program focused on civilian uses (although it is not difficult to convert such technologies from civilian uses into military ones). Those space efforts not closely related to economic development, such as the manned space program, were suspended or cancelled. As in earlier eras, economic constraints presented a continuing problem. However, the average GDP per capita over these ten years rose from below $300, which meant the Chinese people were still struggling to rise above the swamp of poverty. The overall pie of government expenditure was relatively fixed initially; thus it is understandable that the increasing expenditure on other sectors meant that the slice of expenditure on the space program became relatively smaller. As economic growth accelerated, the space proportion of the budget did not rise as a percentage but the total amount did.

The pressures on the space program to become more socially useful meant budget increases became more difficult to acquire since evidence of progress or improvements became more difficult to achieve. China had got beyond the "gee whizz" stage of space development. That is, China no longer focused on firsts for their program but improvements. The early satellites had immense social and economic impact. Their successors could be more skeptically judged as simply more of the same. Often in evaluating space technologies the question is whether they are cutting edge or not, rather than do they work better than their predecessors. The technical leaders of China's space program were placed in a political bind because dramatic improvements became more difficult to demonstrate or explain to their political leaders confronted by myriad demands on their attention and the resources they had available to distribute.

Economic constraints became correlated with technological constraints, since technological improvements cannot be achieved without significant R&D investment. More importantly, the ten years of the Cultural Revolution destroyed the Chinese education and civil research system. Due to lack of sufficient technically trained manpower, constraints obviously existed, which can be clearly seen from the developmental process of the Fung Yan meteorological satellites. Work on the FY series started in 1978, but the first satellite was not launched until 1988. These meteorological satellites did not require breakthrough innovation.

> At its crudest, all that was required was a television camera and a means of transmitting its image directly to the ground. Once this was done, it was then up to meteorologists to interpret the image and make sense of the swirling patterns of clouds, fronts and cyclones.[24]

The United States launched its first meteorological satellite early in 1960, only two years after its first satellite. But even with eleven years' work, Chinese scientists and engineers still could not design a satisfactory system of gyros to orient their spacecraft. Eventually this problem was solved by buying available American equipment which also allowed for reverse engineering. Moreover, the life spans of the FY-1 series satellites, launched in 1988 and 1990 respectively, were short. The FY-1A only worked for thirty-nine days and FY-1B also had a limited life.

In all, the theme of this era focused on economic development and most space programs were focused on civilian use. R&D activities were mainly focused on those space projects that were directly related to economic development, such as DFH communication satellites, FY meteorological satellites, Ziyan earth resource satellites, and recoverable satellites. China also entered the commercial launch market and recoverable satellite market in order to earn foreign currency. Furthermore, the Chinese government encouraged establishing joint economic ventures along with joint R&D programs with various foreign partners to accelerate the development of civilian-use space technologies. Space activities for military and national prestige enhancement were not given as much priority in this era. Space science and technology applications were set to develop the national economy and improve people's living standards. Being dual-use, however, military capabilities also improved, a win–win strategy for Chinese leaders. For an economically stretched state, this benefit provided additional impetus to their efforts. Separating the military and civilian facets of the program was easy on the surface but in reality both moved forward together – a situation found true for all space participants.

The fourth era (1986 to the present)

The political instrument of transition from the third era to the fourth era came in the form of the 863 plan. In March 1986, four Chinese scientists, including Wang Dayan, one of the twenty-three prominent scientists who had made special contributions to the projects of "two bombs and satellites," suggested to Deng Xiaoping that China should invest in high-tech R&D to keep track of the advanced technologies of the world. This was in line with Deng's view that science and technology was the "primary productive force."[25] As early as 1975, he had pointed out that "priority should be given to scientific research."[26] According to Deng,

> the key to the Four Modernizations is the modernization of science and technology. Without modern science and technology, it is impossible to build modern agriculture, modern industry or modern national defense. Without the rapid development of science and technology, there can be no rapid development of the economy.[27]

Deng also held that science and technology achievements "demonstrate a nation's abilities and are a sign of its level of prosperity and development." "If it were not for the atomic bomb, the hydrogen bomb and the satellites we have launched since the 1960s, China would not have its present international standing as a great, influential country."[28] Those accomplishments, however, were in the past; the future demanded that China be prepared to confront the technological challenges of the future.

The US Strategic Defense Initiative (SDI) and the European Eureka Plan reinforced Deng's belief that if China did not participate in high-tech R&D at the beginning, it would become very difficult to catch up later on. SDI was a ballistic missile defense program announced by US President Ronald Reagan in March 1983. Its avowed purpose was to provide a complete defense of the United States against a Soviet missile attack.[29] The effect would be to render the United States immune to Soviet nuclear coercion. An ancillary effect would be to completely neutralize China's ballistic missile force, which was significantly smaller and less capable than the Soviet arsenal. For China, this was a wakeup call that further advances were essential. SDI as originally configured never materialized; its progeny is a much smaller land-based program first deployed in 2004.[30]

The EUREKA project launched in 1985 was another wakeup call for Chinese policy makers. Initiated by French President Mitterrand, the EUREKA project was a pan-European network aiming to promote collaboration and innovation in high-tech fields including medicine and biotechnology, communication, energy, environment, information technology, lasers, new materials, robotics and transport. Its aim was to strengthen European competitiveness in the global market and raise the productivity of Europe's industries. It became clear that if China did not develop its own high-tech capabilities, it would again be left behind by Western countries. China's one-hundred-year experience as a semi-colonial society made the leaders alert to the possibility of being left behind internationally. However, their interest in advancement was always tempered by concern that their personal political position should not be undermined.

For these and other reasons, this high-tech proposal was supported by Deng, and the government launched the National High Technology Research and Development Program (the 863 program) in 1986 with the aim of enhancing China's international competitiveness and improving its overall position in high-tech R&D across the board. This program incorporates eight priority areas: information technology, space technology, lasers, biotechnology, automation, new materials, energy technology, and marine technology.

As a consequence, the pace of the Chinese space program picked up significantly. In March 1986, the State Council approved the Report on Accelerating the Development of Space Technology, which affirmed accelerated R&D on the DFW-3 communication satellite, the FY-2 meteorological satellite and the ZY-1, the earth resource satellite, and the FWS-2 series. The 863 program was obviously much broader than simply the space effort but the space component came to greater public attention because of the manned space component.

Deng's active support of the 863 program reflected his growing awareness of China's continued technological needs if it were to truly compete internationally. In addition, Deng and his associates became more confident in his ability to control the rate of political change that accompanied the great economic and social changes underway. That confidence was shaken somewhat at the time of the Tiananmen Square incident, running from April to June 1989, but after a period of domestic unrest, the political problems largely subsided while the rate of economic and social change accelerated. The government strategy was to sustain high economic growth as a tradeoff for political rights, a bargain acceptable to many.

Expanding space activities

Figure 4.1 illustrates the increasing activity within the Chinese space program during this era compared to the past. More and different types of satellites were constructed and successfully launched. In the third era, the emphasis was upon recovery from the excesses of the Cultural Revolution while putting in place the infrastructure necessary to continue pushing forward. The fourth era sees China beginning to reach the stage of what might be characterized as a "normal" space program. A normal space program is one where the basic resources are assumed to be in place, both technologically and scientifically, and the question becomes which specific applications or objectives should be pursued. The span of activities pursued expands as quickly as resources, including personnel, are made available. For China, those decisions are still dominated by the political leadership which has become ever more sophisticated in its judgments.

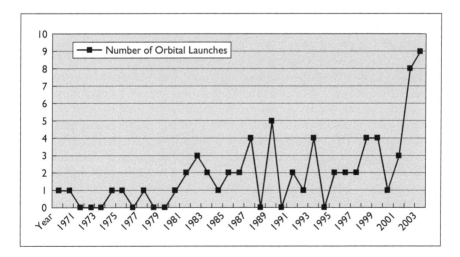

Figure 4.1 Orbital launches by China

The decisions thus far have been in the direction of space program expansion – a trend fitting the leadership's view of China's rising role in the world. The difference is that they are now confident that China possesses the capabilities for success. As will be seen in Chapter 5, Chinese leaders, both political and in the space program, have moved cautiously with regard to human spaceflight. This reflects their awareness that any failure would be both public and damaging to China's position in the world. That said, China has aggressively pursued a growing number of space applications.

This development pattern tracks that of the earlier space participants, with one major difference. The United States and Europe incorporate both public and private (usually corporate) decision makers into the debate over their space programs' future directions regarding development of new space applications. The private sector's focus in both cases is upon economic profitability rather than government prestige. The Russians like the Chinese are confronting the question of how commercial criteria get incorporated into their space applications decisions which are still made by government policy makers.

China at this stage placed greater emphasis on research into dual-use remote sensing satellites and navigation satellites. Three FY meteorological satellites were constructed and launched. Two ZY (earth resource) series satellites were constructed. The ZY-1A and ZY-1B were a joint project with Brazil, used for environment monitoring and natural resources detection. The CBERS program represented an important step in that China embarked on a cooperative program with a non-Asian state – one that was less capable spacewise than China. Such cooperative activity over a lengthy time frame was one manifestation of China joining the international space community as a significant partner. This success made China more attractive as a partner for other states, especially the Europeans. This latter relationship if sustained can generate significant income and advances in technology for China. Saying you are interested in cooperative activity is not the same as actually doing it.

The ZY-2 was developed indigenously and was launched in September 2000. This satellite is evaluated by Western intelligence sources as a military photo-reconnaissance satellite belonging to the Jianbing (Bing means army in Chinese) series, although China's official sources claimed it was for civilian use. ZY-2 was put in a lower orbit than ZY-1 and technologically was more mature than ZY-1 with a resolution of two meters or less.

Three series of FWS satellites – the FWS-1, FWS-2 and FWS-3 series – were ultimately constructed during this era. The FWS-1 series consisted of five satellites, used mainly for civilian microgravity experiments. The first satellite was placed into orbit in 1987 and carried experiments on algae growth and gallium arsenide crystals; the second (in 1988) carried three German experiments. Guinea pigs and plants were carried on the third flight (in 1990), while the fourth one (in 1992) carried microgravity experiments investigating the growth of rice, tomatoes, wheat and asparagus. The recovery of the fifth satellite failed in 1993.

The two satellites in the FWS-2 series were launched and recovered in 1992 and 1994 with their missions involving microgravity experiments. The flights primarily tested improvements in the spacecraft. Compared to the FWS-1 series, the FWS-2 satellite series carried heavier payloads and stayed in orbit longer.[31] The FWS-2-1 (1992) focused on remote sensing and microgravity experiments examining the growth of cadmium, mercury, tellurium and protein crystals. The FWS-2-2 carried rice, watermelon and sesame seeds and a variety of animals. Two other FSW-2-2 series satellites were launched and three FSW-3 series satellites were launched by May 2005. All were employed for reconnaissance, earth observation and microgravity experiments. Besides these, seven science/technology satellites were constructed and launched, including one built by Tschinghua University.

This last event signals that China's technological and personnel base has grown to the point where the capacity to build such a satellite can be found outside the usual locations. Through such hands-on activities, valuable experience can be obtained in wider segments of China's S&T community. The Chinese space program has reached the takeoff point where, with proper investment and sound administration, greater technological feats can be completed on an almost routine basis.

According to the tenth Five Year Plan (2001 to 2005), China was to design, develop and launch thirty-five or more satellites, which would be used in satellite-based direct broadcasting and large capacity communications, weather forecasting, oceanic observation, navigation and positioning, disaster mitigation, and seed breeding. The primary thrust of the plan was to foster China's economic development, even though the dual-use aspects remained clear. More importantly, space applications are clearly becoming more routine and incrementally improved.

To meet the demands of the communication satellite launch market, two types of launchers – the CZ-3A and the CZ-3B – were put in operation. The maiden flight of CZ-3A occurred in 1994. Development of the CZ-3B launch vehicle started in 1986, but it was not launched until 1996. Other launch vehicles were also developed in order to expand China's capacity for space launch. The CZ-4's mission profile involved polar sun-synchronous launching with its first flight, made in 1988, involving launching meteorological satellites. Development of the CZ-2E, a two-stage liquid propellant launch vehicle with four boosters strapped on to the first stage, began earlier in 1986. This launcher was first introduced into the world launch services market after its successful test flight in July 1990. The CZ-2E offers a large lifting capability (low earth orbit: 9,500 kg; and geosynchronous equatorial orbit: 3,500 kg). This version meets the market requirements for low earth orbit (LEO) launching capability. In addition, the CZ-2F was built for launching manned spacecraft. Also, in November 2000,

China initiated development of a solid fuel launcher. This new type of launcher was designed as a mobile, partially reusable small launcher and was named Explorer-1 (Kaituozhe-1). Its third-stage engine was successfully tested in February 2000. Subsequent test flights encountered difficulties but work continues on the new launcher.

Design and development of a new generation of Long March launchers was also included in the tenth Five Year Plan. The goal is to produce non-toxic, non-polluting, and low-cost rockets with dramatically better performance, higher reliability, and greater lift capacity. In a presentation in April 2006, the heavy lifter was to lift 25 tonnes or 25,000 kg to LEO and 14 tonnes or 14,000 kg to GEO with its first launch, projected for 2011.[32] This would jump China into the class of the Ariane 5, the Proton in its various configurations, the Delta 4 Heavy, and the Atlas 5 Heavy, the world's largest launch vehicles.

The capstone of the Chinese space program was to be its manned space program (discussed in Chapter 5). All of the technology developments described above in different ways supported that effort. When China sent its first astronaut (taikonaut) into orbit on October 15, 2003, there was little surprise at its success given these earlier technology developments. China has reached a new level of sophistication and steadiness of purpose which makes its success more likely.

Commercial ups and downs

China's decision to enter the commercial launch market in 1985, including use of their recoverable payload system, represented a major policy change for the national leadership. Their gamble, of which space activities represented only one part, was that the leadership could transition China's economy into the global marketplace while sustaining their political control. Operating in the world marketplace brought China's space program clearly into public focus and under intensified scrutiny by a new set of eyes, the commercial sector, whose demands were different and possibly more intrusive. First, the payload owners for comparative purposes were going to need technical characteristics in order to assess which launch system should be engaged to launch their payload. China now operated on the world stage where competitors would challenge China's launch vehicles as less reliable than their own or not compatible with the payload. The China Great Wall Industry Corporation was established in 1980 to handle any commercial transactions.

Second, China's launch vehicles would be carefully monitored by the international space insurance industry. Whatever launch failures occurred could prove detrimental to China's ability to obtain contracts since the payload owners were not likely to "fly naked" – that is, without flight insurance coverage, Since payloads can run into the hundreds of millions of dollars in value, few companies could afford the loss if a flight failure occurred. This reliability question was especially an issue for new startup companies in their first early launches when failures or partial failures are frequent. A launch failure does not have to involve a ball of fire; it can simply be a premature engine cutoff, leaving the payload short of its designated orbit. This happens even to experienced rocket builders, as occurred during a Proton launch in 1997 despite its 92 percent success rate over a forty-year period. China's rockets were well tested but that was domestic; now they moved into a more unforgiving environment – the global commercial marketplace. Despite China's flight experience, in a sense, they started over again, gaining international commercial experience.

Third, the Chinese launchers had been produced under nonmarket conditions, meaning that the economic cost of a Chinese launcher was unknown, especially to outsiders and competitors. China was coming to the marketplace long after the others so arguments that their launchers were underpriced were difficult to refute. In fact, China's launchers did not differ from the original Western (primarily American) lift vehicles which were all immediate derivatives of military launch vehicles, or public programs such as Ariane, where cost considerations were often secondary. This was especially true in the United States where former ballistic missiles were converted into launch vehicles which did not come under commercial control until after the January 1986 space shuttle Challenger accident. Sunk costs of development were written off just as in the Chinese or Russian case. The space shuttle accident broke the earlier US monopoly over Western space launch – a position previously sustained through subsidies and active discouragement of competitors. In fact, the Ariane launcher was Europe's political response to the US launch monopoly which European governments found intolerable. The space shuttle fleet was removed from the commercial market, destroying at one stroke the US attempt to monopolize apace launch of commercial payloads.

Regardless, this price differential immediately created an immense competitive advantage for the Chinese since their prices were significantly lower than those of their Western competitors. The Soviets-Russians also encountered a similar problem when they first proposed bringing their launchers to the marketplace. Both China and Russia were initially forced to accept politically imposed restrictions upon their competitiveness. At that time, the United States effectively held a virtual monopoly over commercial payloads due to the strength of its communication satellite manufacturing sector. Therefore, the United States employed its economic leverage to impose competitive restrictions on both parties.

Quotas were imposed on the number of comsats each could launch and the minimum price that could be charged per launch. Both agreements were limited in duration but their intent was clear, limiting Russian and Chinese competitiveness relative to Western launch companies. After the Challenger accident, the Europeans first, but followed eventually by the Russians and Chinese, moved aggressively to compete for market share in the launch market. The Europeans through Arianespace with its Ariane 4 lifters immediately seized the high ground; their market rose to nearly 80 percent of total launches by 1990 before the others began to successfully compete. The entry of Russia and China clearly reflected the demise of the Cold War and the early beginnings of normalization of space economics. In the short term, political limitations in the form of launch quotas reduced the ability of the Soviets/Russians and Chinese to compete since the implication was that their rockets were cheaper but technically suspect.

The quotas were cast in terms of number of launches permitted of Western comsats and the prices charged. The latter provisions prohibited pricing more than 15 percent below comparable Western prices. Ironically, all of the launch competitors involved were government-subsidized and -controlled; the crucial issue became the cost of living differentials and amount of subsidy permitted. Regardless, this enhanced international competition eventually reached the point where the United States feared being effectively excluded from the now international space transportation market. The mighty fell quickly once the market truly opened up. Upgraded 1950s and 1960s US technologies did not measure up, especially with their high costs. Both agreements had lapsed by 1994–1995 and, with the abrupt worldwide decline in the projected number of payloads, were not renewed.

Ironically, these quota agreements created economic conflicts when US launch companies and satellite builders publicly clashed over the artificial pricing imposed by the quota agreements. Despite the quotas, however, the US space launch fleet proved unable to compete, although the big winner was the Europeans with their Ariane 4 lifter. Within the United States, any apparent conflict disappeared in the late 1990s when the space consolidation process left two major players, Lockheed Martin and Boeing, who now offered both launch services and satellite construction. For China, the ramification of this American business conflict was that Chinese launch activities were carefully monitored both by governments and by competitors.

The global launch market collapsed in the late 1990s because of the delay or cancellation of several proposed multi-satellite communication systems. That decline was not foreseen when the Chinese and Russians entered the market in the late 1980s. The market collapse occurred when, instead of launching 840 small comsats for Telesdeic, none were launched because that venture failed. Other large-scale LEO comsat systems disappeared prior to construction, "paper satellites." The Iridium system with its sixty-six comsats plus six in-orbit spares was launched but almost immediately entered bankruptcy. The result was too many launch systems chasing a declining number of payloads, as the Japanese H-2 and the American Delta 4 and Atlas 5 came on the market. In addition, the Ariane 5 entered service to replace the Ariane 4.

Table 4.1 China's commercial launches

Years	Number of launches	Failures
1985–1990	4	0
1991–1995	8	1
1996–2000	15	2
2001–2005	2	0

Source: China Great Wall Industry Corporation, "International Customers," http://www.cgwic.com/launch/customers.html (accessed February 18, 2006).

In addition, China and its potential customers had to work out how to interact with each other in this new environment. National security concerns did not immediately disappear even as the Cold War dragged to a conclusion with the final collapse of the Soviet Union in 1991. The Chinese also had to learn what customers demanded and work to satisfy those demands if sales were to occur. For the payload owners, a certain comfort level had to be achieved before contracts would be signed. This process took time, which further delayed China's entry into the market as a major competitor. But, as can be seen from Table 4.2, China achieved increasing success until several events intervened.

Some time delays occurred before major contracts were finalized although China had some initial successes from 1985 to 1995. The commercial satellite business at that point worked on long-lead-time contracts, necessary since satellite construction took several years, meaning several years between contract signing and actual launch. Twelve commercial launches occurred with three flying piggy-back with a Chinese payload. A failure, however, occurred on December 21, 1992 when the Optus-B2 (an Australian comsat) was lost, raising initial questions as to Chinese launcher reliability. The reliability question grew even larger when, within a two-year window, 1995–1996, two separate commercial launches failed – both with reported loss of life when booster and payload parts and fuel landed in nearby villages. In addition, a third launch ended up short of the desired orbit when the third stage shut down early.[33] The failures involved large international players, including Intelsat, the global giant with regard to comsat operations. Its Intelsat 708 comsat was lost during liftoff. The other failures involved either joint ventures between Chinese agents and international partners, APsat, or the dominant domestic player in Chinese telecommunications, ChinaSat. As one can see, China's number of international commercial launches fell precipitously, reflecting increased wariness by potential customers plus the general drop-off in the global launch market. Chinese rockets were now suspected of being failure-prone until questions concerning their reliability were resolved. Explaining these flight failures became an absolute necessity if China was to compete globally.

One must note that launch failures are not unique to the Chinese. According to one analysis, launch failures occur at about a 7 percent rate globally. New launch vehicles encounter a much higher loss rate, up to 33 percent on average for first launches by experienced companies, while new ventures encounter a 73 percent probability of failure on first flight attempt.[34] China was not a new participant but, for the international business community, China's status was that of a new start-up.

China now confronted the harsh realities of the marketplace which had to be addressed before insurers would write future contracts covering Chinese Long March launchers; the accidents would have to be thoroughly investigated. Investigations by Chinese officials were initiated, with US satellite builders actively participating in the proceedings, which in itself exhibited a new degree of openness. As part of that investigative process, technical information was exchanged regarding the various technologies involved. This exchange represented a challenge for both sides – the Chinese had to open up their records and data to outsiders, while the American companies had to provide more detailed information regarding their payloads. The investigations proceeded forward with information being exchanged. Meanwhile, congressional concerns arose in the United States that important proprietary information was being released to China. The information being exchanged, it was alleged, incorporated clear dual-use capabilities. In effect, the information received was enhancing Chinese military capabilities.

For those concerned about China's rise to political and military prominence, this incident provided a window of opportunity. The US House of Representatives authorized an investigation, chaired by Representative Christopher Cox, into the accident investigation process with the intent of discovering whether US secrets were being given to the Chinese that would enhance their military space capabilities. In January 1998, the Cox Committee reported back that the Chinese had been given such secrets and recommended that the United States restrict such technology transfers in the future.[35] On a larger scale, the Committee in its report asserted that China was engaged in a systematic effort to obtain US technology secrets, a particularly potent appeal in an election year. The political argument made by the Republicans was that the Clinton administration and the Department of Commerce were so focused on expanding trade relations that this dire threat to American security was ignored.

In 1998, severe restrictions were placed by the US Congress on any technology transfer exports that might potentially threaten the United States. This was done through the simple expedient of having the US Department of State replace the US Department of Commerce as the agency responsible for approving export licenses for the sale and export of technologies considered useful for military purposes.[36] The latter's mission is fostering business domestically and internationally while the US Department of State protects the US national interest especially its security aspect.

The United States has a well established system of export licenses, the International Traffic in Arms Regulations (ITAR), which are adhered to in order to ship US technologies overseas and these now explicitly include satellite payloads.[37] Existing launch contracts using Chinese lifters were effectively suspended while a more intense and skeptical scrutiny of any export licenses occurred. For China, the immediate effect was to halt its ability to compete for future US payloads, never mind losing already signed deals. The other immediate effect, ironically, was to severely cripple American corporations' ability to compete in the global market – the export license approval process became too arduous and lengthy. Indeed, initially the restrictions were enforced rather arbitrarily. For example, Canada and NATO allies were on the original list of suspect states – a situation embarrassing to the United States and totally disruptive of its efforts to compete globally regarding space technologies. Ironically, the effect was to make Chinese technologies more competitive since a number of potential international customers could now access American products only with great difficulty and expense.

For China, the restrictions on its ability to fly US payloads were a major setback but less devastating that it might have been otherwise. The reason is simple, the international launch market went into general decline (as indicated below) which meant that China could upgrade its launch technologies and not be at a severe competitive disadvantage. In addition, China could use this interval to reestablish the reliability of its launch vehicles by successfully flying its own payloads.

In Table 4.2, the obvious decline in commercial launch payloads globally can be seen. The analysis by the US Office of Commercial Space Transportation counts only those payloads available for international

Table 4.2 Global orbital launch activity (total/commercial)

State	1997	1998	1999	2000	2001	2002	2003	2004	2005
China	6/3	6/4	4/1	5/0	1/0	5/0	7/0	8/0	5/0
US	38/14	36/17	31/13	28/7	22/3	17/5	23/5	16/6	12/1
Russia	29/7	24/5	28/13	38/13	23/3	25/8	21/5	22/5	26/8
Europe	12/11	11/9	10/9	12/12	8/8	12/10	4/4	3/1	5/5
Multinational*	0	0	2/1	3/3	2/2	1/1	3/3	3/3	4/4
Japan	2/0	2/0	1/0	1/0	1/0	3/0	3/0	0	2/0
India	1/0	0	1/0	0	2/0	1/0	2/0	1/0	2/0
Total	89/37	82/36	78/36	85/35	59/16	65/24	63/17	54/15	55/18

* Multinational signifies Sea Launch, a consortium of Boeing and others.

Source: Data compiled by the Office of Commercial Space Transportation, Federal Aviation Administration and reported in annual reports, *Commercial Space Transportation: Year in Review* (various years 1997–2005); tables within the report accessible on line at http://ast.faa.gov/. Totals may not add up due to deletion of single launch events in different years by different states.

competition. Those payloads explicitly restricted to national flag carriers are not counted. What is of interest is the general decline in total number of launches and a concomitant decline in commercial payload launches. No participant country was immune to the general decline in comsat launches which blighted the field from 2001 onward.

China also continued pursuing the sale of payload services using its FWS recoverable satellites. A French payload flew in 1987 to kick off China's entry into the commercial sector within this more specialized sector of space activities. The Europeans have proven to be the most interested in utilizing such services. They have consistently used China and Russia as an alternative to the Americans, who have over the years adversely impacted their space efforts especially with regard to space launch. When Russia and China opened up to collaborations with outsiders, the Europeans were receptive.

Concurrent with these Chinese-based commercial efforts, China initiated the process by which it could more easily access Western technologies, not to own in the sense of physical possession, but to be able to access for their operations. These efforts came in the form of joint ventures, primarily involving communication satellites.[38] Several joint ventures were established including AsiaSat in 1988, APStar in 1992, and EuraSpace in 1994. AsiaSat drew its partners from China, Hong Kong (then a British colony) and Great Britain, while APStar was a partnership of three Chinese companies with a Thai company. The latter consolidated several actors within the Chinese communications sector into a Chinese partner called Sinosat with the German company DASA. This venture was interesting because it involved competition with ChinaSat which, along with the Ministry of Post and Telecommunications, dominates the Chinese comsat market. ChinaSat was a government-controlled corporation which handled both military and civilian-commercial satellites. The corporation had been established to manage China's comsats regardless of purpose – another example of the dual-use properties of space technologies in action.

Competition now took on broader connotations beyond merely competing with international competitors; economic and political rivalries were beginning to grow within the Chinese communications industry itself. The state-centric economic model was generally under attack as others rose up to compete with those entities created by the state. The actual rivals were still often other state entities but the competition was real as the government now considered allowing economic failures to occur. For the space sector, the discipline of the marketplace was becoming real. In the end, potential failures were often cushioned or ameliorated because of concerns for maintaining employment. High unemployment might lead to social unrest and political disruption. Failures in state-run enterprises, however, were not universal since the government was struggling to reduce national dissension through sustaining economic prosperity. For the leadership, a delicate balance had to be maintained.

China's space program publicly focused upon the civilian side of the dual-use equation but the program also continued to have a military-political side that never disappeared. In this next section, the question of China's military space ambitions will be discussed. The pressure to engage in military space activities is intense given the US push for such capabilities plus the active disputes over several issues, especially Taiwan. China's response is hampered by an awareness of the gap between the United States and China regarding existing military capabilities.

Military space demands

Since the end of the 1980s, the biggest security concern for Beijing has been Taiwan declaring its independence under the possible auspices of the United States and China having to face the more powerful American military force in the waters of the Taiwan Straits. Taiwan's trend toward independence has become clearer with the election of Chen Shuibian, and some Taiwanese leaders have openly challenged the "one China principle." The Taiwan issue was given special attention in the White Paper released by the Information Office of the State Council, "China's National Defense, 2004":

> The situation in the relations between the two sides of the Taiwan Straits is grim. The Taiwan authorities under Chen Shui-bian have recklessly challenged the status quo that both sides of the Straits belong to one and the same China, and markedly escalated the "Taiwan independence" activities designed to split China. Incessantly trumpeting their separatist claim of "one country on each side," they use referendum to engage in the separatist activities aimed at "Taiwan independence," incite hostility among the people on the island toward the mainland, and purchase large amounts of offensive weapons and equipment. They have not given up their attempt at "Taiwan independence" through the formulation of a so-called "new constitution for Taiwan." They are still waiting for the opportune moment to engineer a major "Taiwan independence" incident through the so-called "constitutional reform." The separatist activities of the "Taiwan independence" forces have increasingly become the biggest immediate threat to China's sovereignty and territorial integrity as well as peace and stability on both sides of the Taiwan Straits and the Asia-Pacific region as a whole.[39]

Since the Cold War, Taiwan has often been a source of conflict between the United States and China. The US commitment to Taiwan has waxed and waned over the years, with the Bush administration clearly supportive of the new democracy, a direct contrast to the mainland with its authoritarian government. After the aircraft collision which happened on April 1,

2001, the Bush administration decided to sell Taiwan weapons worth approximately 4 billion US dollars.[40] That was the largest arms sale to Taiwan since 1992 and China perceived it as a direct challenge to Chinese sovereignty and territorial integrity. In response, China in recent years has bought several Russian submarines to boost its ability to blockade Taiwan. These Kilo class submarines will be equipped with an anti-ship missile system with a range of 140 miles. But China needs to build an ocean surveillance system in order to use the weapon properly. According to the *Washington Post*, "China still cannot find ships at sea," since China does not have a real-time photo-reconnaissance capability.[41] The ZY earth satellite technology might be useful for helping China to develop military reconnaissance satellites. Creating a credible deterrent to prevent Taiwan's independence and possible American military involvement has become the biggest incentive for China to modernize its military. In this environment, the space program got favorable treatment as a military force enhancer. Enhancing military efficiency and lethality is the major role space technologies fill on the battlefield. Weaponization of space has not occurred yet.

Besides Taiwan, several other issues also provoke China's security concerns. The United States–China relationship has fluctuated over the years since 1989. Although the bilateral relationship improved after September 11, 2001, tension existed for thirteen years, from 1989 to 2001, reinforcing China's sense of insecurity, as a result of a series of events: the bombing of the Chinese embassy in Belgrade, allegations of Chinese espionage at US national laboratories, human rights violation accusations, charges of successive violations of non-proliferation commitments, and the collision of the American EP-3 plane with the Chinese F-8 fighter over the South China Sea. The United States clearly viewed China as a potential threat to American security. According to the Quadrennial Defense Review Report (released on September 30, 2001), "the possibility exists that a military competitor with a formidable resource base will emerge in the region [Asia]. The East Asia littoral – running from the Bay of Bengal to the Sea of Japan – represents a particularly challenging area."[42] Meanwhile, the Bush administration's abandoning of the Anti-Ballistic Missile Treaty signed earlier in 1972 has made China uneasy, although American officials claim that the US missile defense system will not threaten China. For China, the missile defense system greatly enhances the credibility of any first nuclear strike by the United States, and the United States might extend the missile defense system to protect Taiwan or further intervene into other Chinese internal issues, such as Tibet.[43]

In addition, "American military action in the Balkans suggested that neither prospective adversaries nor international organizations would pose much of a constraint on US decisions about where and when to act abroad, and US missile defense plans reinforced this worry."[44] Annual reports by the Pentagon identify China as a potentially hostile power against which the

United States needs to be prepared to take action. The ballistic missile defense program (BMD) is aimed at rogue states such as North Korea and Iran but the reality is that China's ICBM forces are small enough to be defeated by a BMD system that works. The question of whether the BMD system works or not is not totally clear.[45]

Moreover, Japan and India are also considered as potential rivals to the Chinese national interest. "The Sino-Japanese relationship is much more complex than any other bilateral relationships of the two countries."[46] On the one hand, these two countries are big trade partners and they are integrated economically; on the other hand, they do not trust each other politically. The territorial disputes over Diaoyu Island and other off-shore oil areas, and the fact that Japan shows a lack of remorse for its actions in World War II, accentuate the distrust. Anti-Japanese sentiment has increased in recent years. In "China's National Defense, 2004," Japan is listed as a security concern. "Japan is stepping up its constitutional overhaul, adjusting its military and security policies and developing the missile defense system for future deployment. It has also markedly increased military activities abroad." These actions are perceived as an overt threat to China's security.

"India's looming nuclear potential has had a substantial impact on China's national security."[47] A closer relationship between the United States and India has alarmed China. China suspects that the United States wants to use the India card to deter China and India is eager to increase its influence in Asia with American support.[48] Furthermore, the strategic position and rich oil and natural gas reserves of the South China Sea are inflaming the territorial disputes over this region. The Philippines, Indonesia and Vietnam have all claimed some ownership of islands in the South China Sea which are also claimed by China. Clashes for example have occurred over the Sprately Islands.

As of 2002, much of China's military equipment was obsolete; command, control and communication capabilities were poor. In the imbroglio of 1996, many people even believed that Taiwan could defeat any possible mainland offense without US military intervention. China's ability to engage a serious fight in the South China Sea is considered doubtful. China's military weakness lies in its ignorance of military construction from 1978.

> When Deng Xiaoping instituted the "Four Modernizations" of agriculture, industry, science and technology, and national defense as China's core development programs in 1978, national defense was the fourth priority and remains in that position. Although the military budget is increasing, it is still inadequate for building a military that can satisfy Beijing's doctrinal requirements.[49]

Figure 4.2 indicates the inadequacy of Chinese military expenditure even after several years' continuous growth. The Chinese expenditure lagged behind

114 Accelerating the rise of China's space program

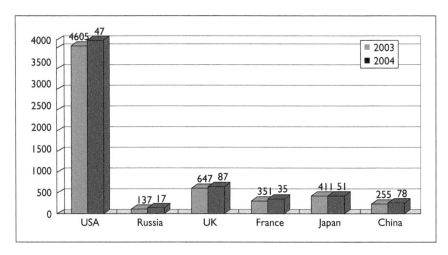

Figure 4.2 Expenditure on national defense (in billion US dollars)

Source: "China's National Defense, 2004," Xinhua News Agency, http://news.xinhuanet.com/mil/2004-12/27/content_2384731_4.htm (accessed May 22, 2006).

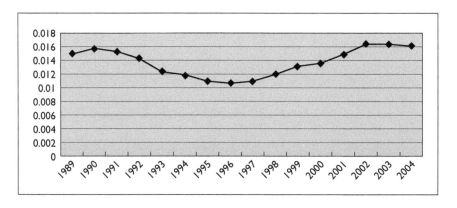

Figure 4.3 Ratio of China's Expenditure on Defense to GDP

Source: National Bureau of Statistics of China.

its regional rival Japan, even taking into account exchange rates. The absolute total military expenditure of Japan was 3.3 times that of China in 2000, a fact which has caused China great concern. China has identified this gap and has begun increasing its military expenditure in recent years to modernize its military force. Figure 4.3 captures this trend. The United States after September 11, 2001 dramatically raised its defense expenditures.

Space support systems are expected to play a more important role than previously.

Over the last 30 years or so, the use of space for military purposes has steadily increased. ... Space is no longer seen as a peripheral adjunct to military planning and operations, but rather is now being more generally regarded as an integral part – a core element of military activity.[50]

The utility of space applications for military purposes has been demonstrated during recent international conflicts, such as the Gulf War, and the air wars in the former Yugoslavia and Afghanistan. Satellites greatly enhanced the communication, navigation and reconnaissance/surveillance ability of multinational forces.[51] Furthermore "even the military space program has had unexpected economic benefits. Military space programs reduce other military costs. Remote sensing espionage from space, in many instances, can be more accurate, more inclusive, and cheaper than air or land-based espionage." Because of these benefits, "disabling the more powerful navy by attacking its space-based communications and surveillance systems and even attacking naval units from space" has become a well-accepted strategy in China.[52] The White Paper on Space Programs released in 2000 suggested the key role of space programs in defense and national security, stressing the need to "protect China's national interests and build up the comprehensive national strength." China has formed a military space research center, the Aerospace Institute of China, whose mission is to study military space technologies and space wars. The long-term aim is that, "by the year 2040 China's space force is set to have become fully operational as an independent service directly under the national military command."[53]

So far China has made space observation, navigation/positioning, and communications the first priority.[54] China is developing the FSW-3 series, a new generation of photo-reconnaissance satellites with 1-meter resolution. According to the *Encyclopedia Astronautica*, ZY-2 (Resource-2), launched in September 2000, was in fact China's first high-resolution military photo-reconnaissance satellite, with a military codename Jianbing-3, although the Xinhua News Agency claimed it was a civilian remote sensing satellite. "The US intelligence sources indicated that it was a satellite for exclusively military purposes, such as targeting missiles at the US and Taiwanese forces."[55]

China has observed that successful military operations cannot be achieved without satellite navigation in modern war. China has launched three navigation satellites named Beidou (Northern Dipper) and its preliminary all-weather satellite navigation system has been completed. Although China did not elucidate regarding its particular military uses, the Taiwan newspaper *China Times* said that these two Beidou satellites belong to the military navigation twin-satellite plan.[56] A third Beidou satellite was launched in 2003 as a backup. Furthermore Chinese overseas space monitoring stations also have dual-use functions, according to the

Christian Science Monitor. They "could be used to track and monitor satellites, rocket launches from Vandenberg Air Force Base in California, and US Navy communications, although nowadays the tracking stations operate as China's own launching monitor."[57]

In the tenth Five Year Plan, China switched from relying upon imported communication satellites to building indigenous communication satellites with larger transponder capacity and extended life spans. Satellite communications technology has been used in national defense. China's first military communications satellite, Feng Huo-1, was launched in January 2000. Feng Huo-1, (FH-1) has both C-band and UHF communications; it is part of the Chinese C^{4}I system (command, control, communications, computers and intelligence), which will enhance the PLA's coordinating and supporting ability.

There are also some rumors regarding Chinese R&D on an anti-satellite weapon, called the "parasitic satellite." According to the Hong Kong newspaper *Sing Tao*, the parasitic satellite will be deployed and tested in space in the near future.[58] No official sources have confirmed this (and, in fact, the evidence appears shaky),[59] but both the tenth Five Year Plan and the White Paper on the Chinese space program mentioned that China would carry out research on space debris, and funding has been allocated since 2000, which may imply anti-satellite weapon construction.[60] The manned space program is mainly to enhance national prestige, but according to the *Annual Report to Congress on the Military Power of the People's Republic of China* (2000), "China's manned space efforts could contribute to improved military space systems in the 2010–2020 time frame. In addition to scientific and technical experiments, Chinese astronauts, for instance, could investigate the utility of manned reconnaissance from space."[61] However, the latter use is not likely since both the United States and the Soviet Union envisioned such activities but found robotic remote sensing both more useful and much less costly.

In addition, the role of satellites in settling international conflicts has become more and more important. The investigation of ethnic cleansing in the Balkans by the International Criminal Tribunal and the resolution of the Ecuador and Peru border dispute were two examples.[62] In recent years, nations involved in the disputes over the South China Sea have used remote satellite data to support their respective arguments.[63] Oceanic satellites and other types of remote sensing satellites are of great significance in dispute settlement talks and in protecting China's interests in the South China Sea.[64]

Other motivations

As in the earlier eras, China employs space activities as one mechanism by which to achieve other goals. What has become obvious is that its activities are under greater public scrutiny, in part because China clearly desires that attention and the state has become too important to be ignored internationally. Those other goals include prestige enhancement (discussed in more detail in Chapter 5 in the context of China's human spaceflight effort), economic development including social progress, and S&T development.

Prestige enhancement

Chinese foreign policy was readjusted in 1996. The new grand strategy is designed to project China as a power that shapes, rather than simply responds to, the international system. To Beijing, the PRC's position will be equal to that of Japan, the EU and Russia in the new world order. China's efforts have been devoted to building a multi-polar world and China is eager to show the outside world that it has the capacity and determination to carry out whatever task it chooses. The manned space program serves as a means to this end. "Though instrumented flight has prestige value, the attention and interest of the world are captured much more by manned flight."[65] Just as the Kennedy administration in 1961 searched for national prestige by calling for a moon landing, President Jiang Zemin in 1989 became devoted to manned space program development for similar reasons. The manned space program's success has made China the third nation in history to launch a manned spacecraft. The program is a three-step program according to *Liberation Daily*.[66] The first step was to launch a manned spacecraft after several test missions. A long-term manned space laboratory will be constructed and launched in the second step. The third step is to build a manned space station similar to the American Skylab and the early Soviet Salyut stations. This program is to be systematic and measured in its pace and intensity. "Failure is not an option," to quote an American official from the early space age.[67]

National prestige provides an obvious incentive for China to engage in a manned space program, as can be seen from official media coverage. After the first successful test flight of the Shenzhou spacecraft, the Xinhua News Agency said that the Shenzhou manned space program would "strengthen the nation's comprehensive national strength, promote the development of science and technology, enhance national prestige, boost the nation's sense of pride and cohesiveness." "The successful test flight demonstrates that China's spacecraft and new carrier rocket are excellent in performance," it said. "China deserves a place in the world in the area of high technology." "This shows that China is fully capable of independently mastering the most advanced technology."[68] Other state television bulletins and newspapers also commented that the test flight was an important step to put the nation of 1.3 billion firmly on the world map and to strengthen the sense of national pride. Note that this aspect only tangentially refers to economic development.

Not surprisingly, the Chinese manned space program won world attention. The *Washington Post* commented that, "Launching space vehicles needs the most sophisticated technology and is a comprehensive demonstration of a country's political, economic, scientific and technical strength."[69] The effort will "boost the nation's sense of pride and cohesiveness [and] arouse the enthusiasm of all ethnic groups." "China's space program is as much about boosting the country's national pride and international stature as it is about scientific research." "In the minds of millions, dramatic space achievements have become today's symbol of tomorrow's scientific and technical supremacy."[70] The political effects are far-reaching. According to *Aviation Week & Space Technology*, China's space program also enhances China's influence in the politically sensitive region. Success will bring further Chinese satellite and booster customers in countries that will continue to have limited technological exchanges with the United States.[71] In the effort of striving to become a major power, the space program, particularly the manned space program during this period, helped change the backward image of the Chinese in the minds of other people and, concurrently, built Chinese national pride and self-confidence. Therefore, national prestige enhancement once again returns as one of the most important motives driving the Chinese space program.

Economic development

In 1970 President Nixon said that,

> We should hasten and expand the practical applications of space technology. The development of earth resources satellites – platforms which can help in such varied tasks as surveying crops, locating mineral deposits and measuring water resources – will enable us to assess our environment and use our resources more effectively. We should continue to pursue other applications of space-related technology in a wide variety of fields, including meteorology, communications, navigation, air traffic control, education and national defense. The very act of reaching into space can help man improve the quality of life on Earth.[72]

After forty years, China is firmly on the same track, since space programs can reduce many costs in the long run. The most obvious example is satellite communications, which are less expensive, compared to ground cable installation.

In China, satellite TV transmissions already cover about 80 percent of the population and TV or distance learning education has benefited nearly 20 million people. The communications satellite has helped overcome difficulties in communications with China's remote areas. The weather satellite system has greatly improved weather forecast accuracy, especially disaster weather forecasts, which has reduced the losses by more than several billion yuan every year. The wider use of satellite remote sensing data has helped solve a variety of problems such as mapping, environmental problems, agriculture and urban planning. Data from the FSW series indicated that the total number of islands off the Chinese coast was 5,000 instead of 3,300 and that Chinese farmland covered 125.3 million hectares rather than the previously believed 104.6 million hectares. Satellite-based railway route selection has saved about 100 million yuan by bypassing a land fault.

The incentives for economic development can be seen in the tenth Five Year Plan. China plans to accelerate R&D on the new DFH comsat series to increase communications capacity, efficiency, bandwidth and life expectancy. R&D on the direct-broadcasting satellite is being carried out and part of the mission of these satellites is to provide high quality television broadcasts and educational and information transmissions to the fast-developing western area.

The launch of the first seed breeding satellite occurred in 2003. Seeds experiments were performed and these space seeds are expected to be suitable for arid western China. The ZY satellite series will be designed with higher resolution and a longer operational lifetime and part of its mission is monitoring natural disasters, desertification and environmental pollution, estimating crop growth, city planning and assessing national resources. The Haiyang ocean remote sensing satellite series is used for mapping oceans and measuring depths and currents, and thus fisheries will benefit.[73] China plans to launch six Feng Yun meteorological satellites by 2008 to provide full coverage for the Olympics held in Beijing. The strong focus of the space program on economic development is even seen in the manned spaceflight program. One hundred kilograms of seeds were carried by Shenzhou I; and space-based life science research was carried out on Shenzhou II and Shenzhou III. The experiments are considered of great significance for producing purer and more effective biological products.

The strong motivation for economic development on the one hand rests on the endogenous demands for economic development. On the other hand, China is facing increased external competition due to its joining the World Trade Organization, and space technology becomes a means for enhancing China's international competitiveness. Establishing a satellite communications network, for example, will have an annual value of 20 billion yuan and about 100 million people in remote areas will benefit.[74] China's potential for satellite telecommunications remains huge. The main-line telephone owning rate of China is 12.1 percent, while that of the United States is 75.6 percent. Satellite mobile telephone users in China will increase to 1.2 million by 2010 from currently 200,000–300,000 users.[75] But, so far, more than 50 percent of China's space communications goes through foreign communications satellites and the competition for occupying satellite orbits is fierce since many entrepreneurs are eager to enter the Chinese market. Besides investments in communications satellite R&D,

Table 4.3 Commercial orbital launch revenues (in current dollars, millions)

State	1997	1998	1999	2000	2001	2002	2003	2004	2005
China	$148	$90	$23	0	0	0	0	0	0
US	923	911	672	370	167	338	304	375	70
Russia	351	313	670	671	178	412	178	290	350
Europe	970	763	750	1,433	948	1.133	526	140	490
Multinational	0	0	85	255	170	75	225	210	280
Japan	0	0	0	0	0	0	0	0	0
India	0	0	0	0	0	0	0	0	0
Total	$2.392 billion	$2.112 billion	$2.294 billion	$2.729 billion	$1.463 billion	$1.958 billion	$1.2 billion	$1.0 billion	$1.2 billion

Source: Data compiled by the Office of Commercial Space Transportation, Federal Aviation Administration and reported in annual reports, *Commercial Space Transportation: Year in Review* (various years 1997–2005); tables within the report accessible on line at http://ast.faa.gov/. Revenues may not add up due to rounding and data adjustments in subsequent years.

other measures have been adopted, such as supporting companies that see gains from economy-of-scale operations. In December 2001, the China Communication & Broadcasting Satellite Company and the China Oriental Satellite Communication Company officially merged to establish the China Satellite Communication Group Corporation.

As indicated earlier, commercial launching is crucial for China's high-tech exports, since compared to other Chinese S&T sectors, commercial launch has the larger international competitive advantage. But, challenges clearly exist in this sector. Table 4.3 clearly lays out the precipitous decline from 1999 onward in Chinese revenues from launch activities. China's revenue from international launches slid to zero. Ironically, the United States likewise declined dramatically in terms of launch revenues. In fact, without US DOD funding, the American launch industry was in effect dead, losing out to the Russians and Europeans. US corporations in 2005 made more money selling commercial launches on Russian-derived vehicles than from their own domestic lifters. China's return to the market is underway as a result of its string of successful launches. Part of the problem is that the Chinese launchers have a much shorter history than those of the United States and Russia, which means that when losses occur, they have a disproportionate impact on their reputation.[76]

As discussed earlier, China is planning to grab a larger slice of the global market by developing a new series of cheaper and more powerful launchers by 2010. Premier Zhu pointed out in his Government Report at the Ninth Session of Congress that the Chinese government would strongly support R&D on new types of launcher so that competitive advantage in some S&T sector can be established.[77] This new series will have high thrust, zero toxicity, zero pollution, low cost, and high reliability. China's official newspaper, *Liberation Daily*, said that these new launchers would meet the demand for the next thirty to fifty years and lift China to the top of the world commercial launch market.[78]

An important facet of Chinese policy is greater interest in the question of environmental protection especially in the context of fostering economic growth and development. "The biggest problem of the next century is to feed, house, educate and sustain at a decent standard of living a world population that will probably reach 10 billion; to avoid ecological collapse; to secure an abundance of resources." China's ecology is brittle. Farmland loss has kept increasing since 1952; as of 1996, the average farmland per person in China was reduced to 0.07 hectares from 0.2 hectares in 1952.[79] More disturbingly, nearly one-third of the arable land is affected by erosion and silting and the consequent topsoil loss has led to 30 million tonnes of erosion of nitrogen and phosphorus every year. Since 1978, 700,000 hectares of land have turned into desert each year, and topsoil from another 16,000 hectares of farmland turns into potential desert. The loss of topsoil threatens Chinese agriculture which is the mainstay for over 70 percent of its rural population. Grassland has also been degraded due to population increases, deforestation and desertification. Forests have lost 79 million hectares over the past forty years and now forests cover only about 12 percent of the total land area, which further accelerates the processes of erosion. In the tenth Five Year Plan, to achieve sustainable development, the China Aerospace Technology Group and the State Environmental Protection Bureau are setting up a network of three small satellites to provide real-time monitoring of the environment and analysis of disasters. Future losses are expected to be minimized with the application of this technology.[80]

Technology development

The space program has finally reached that stage where its impact on the national effort at S&T development is one of enhancing Chinese competitiveness. Earlier, Creola demonstrated the importance of the European space program in that process; he pointed out that,

> if Europe does not step up its space efforts, it will find itself uncompetitive in many high-tech industries, reduced to dependence on the national priorities of others in vital areas such as navigation and unable to have its views taken seriously over issues such as the negotiation of telecommunications slots and rules regarding the limiting of space debris.[81]

China is in the same situation. As in the previous era, one of the motivations for space program development remains promoting overall S&T development. The "Report on the Outline of the Tenth Five Year Plan for

National Economic and Social Development" passed by the Fourth Session of the Ninth National People's Congress held in March 2001 said that government must vigorously promote significant high-tech research to make breakthroughs in crucial technological fields that are closely related to national economy and security.[82] The space program has been chosen as one of those fields. It is one of the seven pillar S&T sectors strongly supported by the government since 1986. The general point is that if China does not make a similar concentrated effort at S&T development, the technological gap between China and the Western countries will widen.

In a fundamental sense, the space program has become one marker of Chinese technological sophistication. As was said before, technological development is closely correlated to economic development. On the one hand, technological development provides the means for sustainable economic development; on the other hand, successful economic development will inevitably facilitate technological development by investing more in the R&D sector. China's science and technology developed fast from 1985 onward due to institutional reforms in the science and technological sectors.[83] China's technological development can be seen in Figure 4.4, which clearly indicates an increasing trend. In 1992 (the data are based on World Bank data, which start from 1992), high-technology exports formed only 6 percent of all manufactured products, but in 2003, this increased to 27 percent. Obviously, China's technology has developed at a fast pace. Ten years ago, the usual kind of Chinese products sold in the world market were labor-intensive goods such as clothes and crafts, but now more and more goods sold in a store such as Best Buy (a leading US retailer) are made in China. Although Best Buy merchandise cannot be categorized as

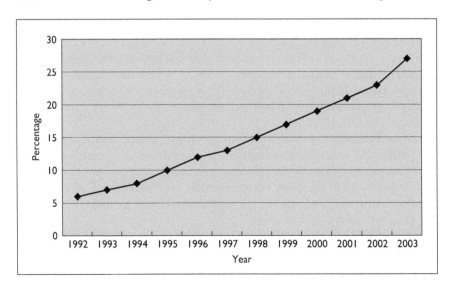

Figure 4.4 China's High-technology Exports (% of Manufactured Exports)

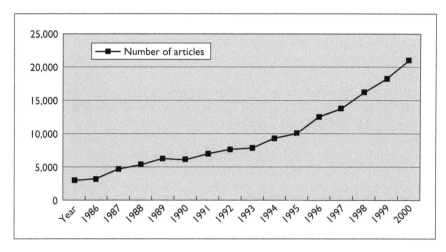

Figure 4.5 Scientific and Technical Journal Articles Published in China

high-tech in a strict sense, nevertheless one must admit that a digital camera sold in Best Buy is more high-tech than the clothes sold in Walmart.

Another indicator of technological development is patent applications. In 1995, the number of resident patent applications within China was 10,066, but in 2002, this rose to 40,346. Chinese patent applications in the United States have also increased dramatically in recent years. In 1994, the number was 100; in 1999 it increased to 257; in 2004 it dramatically increased to 1,665. Meanwhile scientific and technical journal articles also indicate that the technical constraints of the Chinese space programs have shrunk in this era. Figure 4.5 shows the number of Chinese scientific and technical journal articles published between 1986 and 2002 (the data are only available up to 2002). The graph clearly shows a developing pattern. In 1986, the number of articles was only 2,911, but in 2002 this number had increased to 20,978.

Innovation in the science and technology field can be further proved by the number of journal articles published by Chinese nationals in the top academic journals of the world. In 1989, the number was 4,538, but in 2000, the number increased to 16,098, which not only exceeded Indians the number by and Brazilians but also caught up with the Russians. Therefore the technical constraints in this era have shrunk and China has much more credible economic and intellectual resources to mobilize. These resources provide a necessary condition for the success of the Chinese space plans.

Domestic political benefits

Jiang Zemin was promoted to the position of Secretary General by Deng in 1989. Jiang, the former mayor of Shanghai, had no solid political base in the CCP Central Committee and central government. With the support of Deng, he established his political influence by controlling the political center in Beijing. Jiang's political success lies in the fact that he is the political compromise between reformers and hardliners. He inherited his predecessor's political principles and he vigorously promoted economic privatization, but he disapproved of political liberalization. Although the liberal faction lost most of their battles in 1989, their great influence continues. Meanwhile, the hardliners headed by the Maoists strongly oppose any kind of liberalized reform. Conflicts were heated in many policy-making fields. For Jiang, stabilization was the overall priority. He argued that China cannot afford political turbulence any more, since only in a peaceful environment can China concentrate upon its economic construction. Thus, balancing the conflict between the two factions became Jiang's political priority.

The space program ironically serves the needs of both liberal and hardliner factions. To the liberal faction, with intellectuals as its backbone, space program development stresses the importance of knowledge and thus increases their political influence. To the hardliners, space programs are centralized government projects, for which the government makes plans, allocates funds and assesses the final products, a perfect reversal of the previous system in their eyes, although after more than ten years of privatization reform, the left wing of Communist Party is almost silent today.

But new problems appeared and the space program supported elites' political positions. With the reform of state-owned enterprises, unemployment became the most severe transitional problem. According to official figures, employment in state-owned enterprises fell from 112 million in 1996 to 86 million in 2000. Meanwhile, urban collectives reduced employment from 30 million to 17 million. Those massive layoffs make it difficult to successfully job-hunt. Moreover, excessive labor in rural areas cannot be absorbed by urban industrial sectors. Prevalent corruption has aggravated this transitional bitterness and thus further weakened the government's legitimacy. Urban protests about layoffs and rural upheavals have occurred and are ongoing.[84]

Just as with Mao thirty years ago, space programs, particularly the manned space program and lunar landing program, reflected the desire of the political leadership to use the space program as an internal stimulus to appease people's dissatisfaction. According to Chinese officials, the manned spacecraft launching "was intended to unify and strengthen the country." Thus the space program seems a win–win situation, and therefore the government at this stage has made an unprecedented effort to promote it. However, huge investments in the manned space program, in constructing the space lab and the lunar landing, inevitably "crowd out" the resources that should have been for other uses. Thus, space plans made by this administration might eventually become an unacceptable burden for Chinese society under a new government, although the pace might slacken at some point in the future just as it has for the earlier space pioneers.

Economic considerations

The Chinese space program developed rapidly after 1986. This fast development is based on China's fast-growing economy, which has provided a solid material foundation for Chinese space programs. As can be seen in Figure 4.6, the growth of the economy has been the major asset supporting China's space program. As long as that remains strong, political support for the program will continue.

Figure 4.6 indicates China's economic development (1986–2004) as measured by GDP per capita. In 1986, China's GDP per capita was only $288, but in 2004 it was $1,162, about four times that of 1986. Note that China has a population of 1.28 billion, thus the overall economic scale is very big. According to the online CIA world outlook report, "Measured on a purchasing power parity (PPP) basis, China in 2004 stood as the second-largest economy in the world after the US."[85] This huge economy is sufficient for China to fulfill its ambition in the space program sector if we compare China's economy today with that of the former Soviet Union in the 1960s and 1970s. The difference is that China is moving away from the earlier Marxist economic model which means that resources, while plentiful, may not be as accessible for state use as under a planned economy.

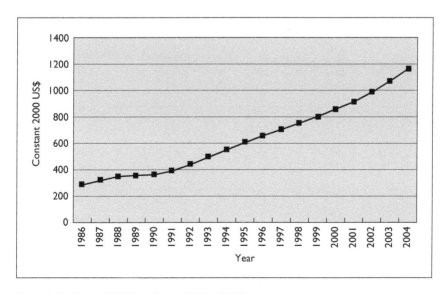

Figure 4.6 China's GDP Per Capita (1986–2004)

Conclusion

China in the third era saw the clearing away of the obstacles to progress in many fields, not least of those being space activities. In the current period, the fourth era, one sees those earlier efforts coming to fruition once the political barriers to success were removed. In Chapter 5, we will backtrack a bit to pick up the thread of the Chinese effort at human spaceflight, to many the epitome of space activity at the highest level.

Chapter 5
The politics of Chinese human spaceflight

Introduction

Since the 1960s, China has widely broadcast its clear intentions to pursue an independent Chinese human spaceflight effort. In this chapter, our analysis focuses upon the political-technological justifications for initiating such a program given the extraordinarily high costs required for engaging in such a task. China in this arena also clearly tracks the earlier human spaceflight efforts of the two Cold War antagonists, the Soviet Union and the United States. China's path to eventual success, however, proved more difficult given its huge technological and economic deficits; at least partially overcoming those deficits first was absolutely essential for success. It is for those reasons that one can speak of China having initiated two distinct human spaceflight programs. The first failed in the cauldron of the Cultural Revolution while the other is ongoing. For both, the motivations were the same but the political circumstances were totally different.

Why human spaceflight?

China's motivations are similar to those often expressed by the two original space pioneers although in this historical context China presently races alone. The motivations are international prestige and security. Space activities become a surrogate for demonstrations of national military strength. China races to catch up with the two leading space powers (the United States and Russia) as it is doing with regard to the Europeans and Japanese in other space applications and space science areas. No other state stands in an equivalent status to China presently, although India expresses similar intentions albeit more long-term in nature. India's political importance comes in its role as a regional rival to China, sharing mutual borders over which armed conflict occurred in 1962 and 1963. China attacked India over a border dispute in the Himalayan mountain range. However, currently, Pakistan is a more immediate security threat for India. India's rivalry with China contains some military overtones but those are not as substantial as those tensions existing potentially between China and the United States and increasingly Japan.

On the other hand, India, for China, provides a benchmark against which to measure its progress from its earlier underdeveloped status, while the Americans and the Russians still remain the measure of what must be accomplished in the space arena if China is to assume the global position its leadership expects as China's due. China's leadership perceives the other two states, Russia and the United States, to be in relative decline compared to China. Their apparent decline, however, does not affect China's interest in successfully conducting human spaceflight.

With regard to human spaceflight, other possible competitors such as Japan, Europe and Canada are presently pursuing different national paths by relying upon Russia and the United States as partners in human spaceflight activities. The Japanese and European space programs across the 1980s through the mid-1990s actively pursued independent human spaceflight capabilities. In both instances, the cost–benefit ratio was found to be seriously deficient as each coped with domestic economic pressures. The announced US withdrawal in January 2004 from the International Space Station (ISS) program around 2016 to 2017, combined with the second shuttle grounding after the February 1, 2003 Columbia accident, has revitalized European interest in seeking independent human spaceflight.

How that might be accomplished is unclear although there are indications that the Russians might lead a cooperative program, the Kliper, a reusable vehicle.[1] Russia has acknowledged its need for outside (i.e. international) funding if the Kliper is to be built quickly to replace the Soyuz. It is furthering cooperation with the Europeans with projected Soyuz 2 launches from Kourou. Japan's efforts remains stymied by domestic budget issues plus a recent reorganization and consolidation of its space efforts. That reorganization further reduced the program's visibility and importance. The reality is that in both cases, the cost–benefit ratio remains politically suspect, although their interest is growing because the United States is moving off in its own direction pursuing its Vision for Space Exploration. The termination of the space shuttle program by 2010, possibly earlier if another shuttle accident occurs, further whets their interest but the cost dimensions remain intimidating. There is no compelling political case for making those expenditures.

Earlier, President Ronald Reagan in the 1980s offered to fly a Chinese astronaut on a space shuttle mission – an offer withdrawn after the space shuttle Challenger accident in January 1986. Around the same time, the Soviet leader, Mikhail Gorbachev, extended to China the opportunity to fly a Chinese cosmonaut to Space Station Mir, but there is no evidence that the Chinese seriously considered this offer; their focus remained on achieving independent human spaceflight. What earlier had been a decision beyond China's immediate control, access to orbit, was now their political choice alone. China wants to come to the table as an equal, something both the Americans and Russians have been reluctant to concede. China therefore waits while it further develops its human spaceflight technologies. Future international cooperative human spaceflight endeavors will be hard-pressed to ignore their efforts.

China's willingness to go it alone meant that their progress was dependent upon their indigenous efforts, a factor that had already delayed their arrival in space. But, that meant when success came, China and, by extension, its leaders received all the glory and acclaim. Attempts to deny their success in the West by attributing any success to the Soviets' (later Russians') assistance were rejected by most observers. Every state at some point has received assistance either directly or indirectly in pursuing space activities. The Soviets went their own way in the 1940s but were cognizant of the Germans' efforts, while the US space program in part was built directly upon the German efforts. In fact, Brian Harvey asserts that the earlier rocket programs in all three states were all originally premised on the German V-2 rocket as the original prototype.[2]

As a result, China by default becomes a beacon for other economically and technologically challenged states in their quest for greater international prestige and defense capabilities plus accelerated technological development. For such states space activities become one critical mechanism for achieving those important national goals, with human spaceflight at the pinnacle. Space policy analysts tend to strongly emphasize the technical aspects alone (which are obviously important) rather than the policy side when analyzing potentially emergent national space participants. That analytic myopia underestimates the political dynamics occurring as economically challenged states strive in this highly visible and prestigious technological arena to leverage themselves higher up in the international space pecking order.

Their motivations are always mixed, as is described in this chapter as China begins aggressively, by its mere existence, challenging existing space policy arrangements. Those arrangements reflect an earlier era, the Cold War with its two-power nuclear arms race paradigm. Adding human spaceflight to its technological repertoire becomes a major advance for China, a clear marker of technological excellence for a state formerly ranked very low in that area. For China, success matches up with its heightened view of its deserved international status. Similarly, on the political level, entering the UN Security Council as a permanent member was an equivalent step in the political realm. In fact, China is now in a position of political superiority relative to its rival, Japan.[3] Japan desires to be made a permanent member of the Security Council but China holds a veto which it is willing to exercise.

Uniqueness of human spaceflight

Human spaceflight represents the pinnacle of space activity for several distinct reasons including both its extreme technical difficulty and great cost but also its uniqueness. Only two states, the United States and the former Soviet Union, now the Russian Federation, had accomplished the feat of successfully launching humans into Earth orbit and returning them safely in a national spacecraft prior to October 15, 2003. There are rumors that China earlier in the 1970s lost a taikonaut but no confirmation was ever received.[4] Overall, the evidence available up to this point further highlights the uniqueness of the feat.

Humans remain fragile creatures so that conducting successful crewed spaceflight involves not only physically launching the payload to orbit (a technologically difficult feat in and of itself) but also insuring the survival and good health of the crew in space and during reentry. The forces and stresses manifested during such launches must be ameliorated, while once in orbit, a sealed safe environment must be provided and maintained against all outside factors, and later a safe return must be insured.[5]

The dangers during liftoff include malfunction of the boosters, dynamic overpressures during early flight, vibration from shock waves during liftoff (the pogo effect), improper stage separation, and navigation failures. Improper welding and faulty wiring can fail during liftoff or subsequently, and debris can damage the booster engines or external areas of the spacecraft, resulting in improper orbits. The crew can be injured or killed by loss of cabin integrity during liftoff and while in space. Spacecraft can be struck by debris or meteorites. Navigation errors can result in a ballistic return to Earth at too steep an angle. Heat shields can fail during reentry, as occurred with the shuttle Columbia – a small breach of the leading edge of the left wing resulted in total vehicle loss. Vehicles can land off-course, creating hazards during recovery, such as landing on water as opposed to land.

The immense difficulties inherent in sending humans to orbit were not truly appreciated by the public for many years. This ignorance resulted from the deliberate opaqueness with which the Soviet space program was conducted, plus the Americans' deliberately fostered image of extreme competence. The very real dangers remained fairly abstract rather than the constant reality they represented. Through secrecy and image management, the two pioneer space participants deliberately downplayed the obvious hazards of human spaceflight. It is worth remembering that much of what the general public knows about space activities comes through fictional stories, whether movies or literature, rather than personal technical knowledge.

Also, for many years, human space flights were particularly unique in their requirement for safe return of the payload sent to orbit earlier. Almost all early space payloads were effectively one-way ventures except for the early intelligence spy satellites, which ejected film canisters, to be retrieved either in the air (catching the parachuting film capsules) or landing on the surface. The American Discover-Corona spy satellite program endured thirteen straight failures before success was achieved.[6] Those failures covered the spectrum from launch failures to loss of film canisters during reentry. Those early recovery efforts were phased out once improved signal transmissions sent sharper imagery from orbit. That simplified the task and extended the life of remote sensing satellites.

Disaster can strike during reentry, as was seen with the Genesis disaster spacecraft in September 2004 when its parachute failed to deploy. Instead of being retrieved in the air during its descent by a helicopter, in a similar way to the early spy satellites, the return capsule slammed into the desert at high speed.[7] A several-year scientific investigation of cosmic dust was literally crushed, although apparently major research components were salvageable.[8]

None of the above situations are easy: crews have been lost during launch (the space shuttle Challenger), almost lost in space (Apollo 13), and lost returning from orbit (the Soviet cosmonaut Vladimir Komarov in 1967 and the space shuttle Columbia's seven-member crew); all of these accidents occurred despite prior flight experience and, in the shuttle case, some warning as to probable flight hazards. The margins for error remain thin and unforgiving even for advanced states. This reality is unfortunately discovered each time a space mishap occurs. For the public especially, such disasters are real shocks because everything appears so easy on the surface; success is routine and expected as a matter of course. Those earlier successes by the pioneers increase the pressures on new entrants into the realm of human spaceflight – early failure must be avoided if at all possible given the deleterious effects upon the prestige sought by such activities.

Other states' nationals, in addition to the Soviets/Russians and Americans, have launched to outer space but those individuals all traveled as passengers of the first two. Both the Soviets and Americans earlier engaged in cooperative programs facilitating such flights by other states' nationals, their putative partners. This tactic was partly pursued in order to maintain control over their allies' independent crewed space efforts. The Soviets sent their guests to their various space stations (Salyut and Mir) while the United States flew them on shuttle missions, with tenures on the ISS expected to follow once shuttles return to flight in 2006. Both states described themselves as having quasi-international programs, in which their space activities became in practice yet another device by which to politically bind their allies more closely in this important field. Thus, all cooperative human spaceflight efforts were premised on the assumption that the space power itself remained the essential component in those efforts.

No subordinate space participant was allowed to operate in the critical pathway for successfully completing the cooperative project. Critical pathway refers to those program components or activities which must be accomplished if the particular program is to be successful. Total failure occurs if one of those activities is not accomplished exactly when needed. The International Space Station (ISS) became more truly international and not just another American-run consortium when both the Russians and Canadians were placed in positions on the critical path for the ISS program. Both participants encountered difficulties but the US could do little to redress them except to support their efforts to overcome problems. For the Russians, the critical component was the Zarya or Service Module, while for Canada, the Canadarm 2 was the critical component necessary in order to complete construction of the ISS. Both states encountered difficulties, especially the Russians financially, with the result that the ISS construction lagged. ISS construction was also further delayed by US shuttle problems.

Since the 1980s, China has been interested in independently launching Chinese nationals to outer space again. Their original space efforts in the 1950s were at first premised upon cooperation with the Soviets, but all of that potential technical assistance evaporated with the political schism between China and the Soviet Union in 1960 (see Chapter 3). As a consequence, China's human spaceflight effort comes without the global fanfare of the original voyagers because of the resulting time lag but still holds enormous symbolic significance since China becomes the first state from the former Third World to successfully pursue their space ambitions so far. India is pursuing that goal albeit at some point in the near future, while Europe and Japan are even farther behind politically in pursuing the race to space. The latter two are technologically closer to success but lack the necessary political impetus.

Human spaceflight demands that a state acquire cutting edge space technologies either through independent national development or through assistance from others. The latter situation has not been the usual pattern because the possession of the requisite launch and other technologies has been deemed a military and economic advantage. Therefore, any assistance comes indirectly, much like dissemination of nuclear technology. The basic principles underlying spaceflight are understood; it is the applications that remain difficult. China's successes therefore demonstrate its capacity to do the very difficult – a feat especially impressive to other economically challenged states. One important ancillary effect is to pressure China's regional neighbors, especially Japan, Taiwan and South Korea, states aligned with the United States on many issues. China has returned to its status as a "big tiger" compared to the "little tigers" along the Western Pacific Rim.

Once the capability for achieving orbit was publicly demonstrated in October 1957, followed by human spaceflight in 1961, other states' engineers and technologists could observe that feat. Knowing that human spaceflight was now truly possible, the technologies can in effect be back-engineered even without a physical copy in one's possession. This possibility exists for all states but their numbers will remain comparatively few simply because the sunk costs are so high. The investment demanded far exceeds the likely benefits. Even economically advanced states have found the economic challenges sufficient to deter their interest on an individual state level.

For developsping societies, human spaceflight may be beyond their capacity but many find space activities important and attempt to participate using other states' launch capabilities. Nigeria and Algeria, for example, operate small remote sensing satellites built and launched by others. For most underdeveloped and developed states, the reality is that human spaceflight for them awaits the arrival of a truly commercial spaceflight sector. China has moved beyond those limitations even though its economic challenges remain significant.

For China, as was stated earlier in Chapter 2, the justifications for conducting human spaceflight are myriad, but some are clearly more important than others especially in its original attempt to develop the requisite flight capabilities. Initially, political considerations dominated – human spaceflight was clearly seen as a marker of technological excellence with obvious military overtones. China was dramatically disadvantaged economically relative to the two superpowers. Space launch became China's initial declaration of its capability to deliver warheads, especially nuclear, over vast distances. Missile launches can be monitored by other states but secrecy usually reigns – orbiting a satellite for the first time is trumpeted to the heavens. Thus, for China, space launch filled the same political-military need that it did for Khrushchev and the Soviets in the 1950s, a clear signal that the United States must consider their (the Soviets', now the Chinese) ability to strike the US homeland. Human spaceflight therefore became a further declaration of China's capabilities because it moved beyond the simple act of launch to include the tracking and eventual recovery of the human payload, indicating even more robust technological capabilities.

In developing its human spaceflight program, China benefits from observing the earlier participants, especially the United States whose program was more open to outside scrutiny. As Joan Johnson-Freese has pointed out, the US Apollo program provided a prototype for China in pursuing economic and technological growth.[9] President Lyndon Johnson in the 1960s consciously employed the lunar landing program as the political vehicle through which the southern United States (the Old Confederacy), a region generally economically behind the other US regions, could be drawn into the national economy by investing funds and facilities in the region. The long-term economic effects were enormous and continued even after the Apollo program shut down in the early 1970s. China likewise is using its investments in the space program generally, but especially the human spaceflight component, to upgrade its technical sectors to compete on the world market.

Chinese nationalism and national aspirations both strongly support this expansive view of China's stature and role in the world. Human spaceflight becomes another very public instrument for securing recognition of China's stature. This is particularly important since in many technology areas China remains a developing, albeit very large, country. The difference between early Chinese justifications for space activities and those presently articulated is that China at first reemployed space in a reactive negative manner: trying to convince others that China would be too difficult to defeat and would inflict excessive damages on the aggressor in the process. Presently, Chinese space activities are being more positively portrayed as emblematic of its rising power and influence. The military undertones regarding China's ability to inflict severe damage in any future military conflict are implicit rather than explicitly stated. So, for expenditure of relatively small amounts, larger political benefits have accrued. The amounts, however, were large for the Chinese when required for investment.

China's human spaceflight program: false starts and success

In January 1956, Mao Zedong declared that science and technology development was now a national priority that must be aggressively pursued if China was to advance and stave off its external enemies. Subsequently, as described in Chapter 3, China embarked on its quest for space launch and missile capabilities, the latter being the more important program at that time. By 1961, with Yuri Gagarin's orbital flight followed by Alan Sheppard's more modest suborbital effort, the Chinese began considering the issue of human spaceflight. The space age was young and China had not yet flown to orbit but there arose a sense of urgency that permeated the space arena. States felt they had to pursue such activities or else be left behind technologically and ultimately economically. The Europeans were the most publicly concerned about this issue which appeared politically in discussions over the growing "technology gap" between Europe and the United States.[10] China's efforts remained much more obscure due to government secrecy and its general political and economic isolation from the flow of world news.

Project 714 – the first attempt

China from the early to mid-1960s discussed the concept of a manned space program, but that initial effort floundered on the irresolvable issues of resources, technology and politics.[11] Given finite resources, its priority remained missile development with space launch clearly more distant and human spaceflight well beyond that. Nuclear weapons technology was also a higher priority than any idea of human spaceflight. Space technology development lagged initially while China completed its first task of creating the infrastructure capable of producing useful flight equipment. That task took time and resources but was severely hampered by politics.

Discussions picked up momentum in 1965 when the Chinese Academy of Sciences (CAS) was charged with developing a long-term space program.

In March 1966, the CAS convened several meetings, with the new Ten Year Plan charging the space sector to focus on the goal of first orbiting a satellite, following that feat with a recoverable satellite and culminating in flying manned spacecraft. Reaching these goals proved more difficult than originally imagined when domestic political events spiraled out of control.

Later in 1966, the Great Proletarian Cultural Revolution kicked off with great disruption of Chinese education, scientific and technical institutions. The planned suborbital flights carrying various animals, including dogs and monkeys, were disrupted and progress halted. The missile and space programs were partially protected by being transferred back under the auspices of the People's Liberation Army (PLA), but progress significantly slowed after Lin Biao's death.

In this early period, the major visible external sign that China's human spaceflight aspirations continued came in April 1968 when the Space Medicine Project Research Institute was established (although its existence remained classified) under PLA control. The Institute survived even after the original manned program was cancelled. Its existence was a symbol of China's continued interest in manned spaceflight even at the lowest points of the Chinese space program. On July 14, 1970, the manned space project, Project 714, was approved and the R&D on spacecraft and the selection of taikonauts (or yuhangyuan) began. This formal approval came after the first orbiting of a Chinese satellite on April 24, 1970.

The new program envisioned Chinese astronauts or taikonauts flying the Shuguang-1 (Dawn) spacecraft to orbit by late 1973.[12] The Shuguang-1 spacecraft consisted of two modules: a reentry capsule and an equipment module. The reentry capsule was capable of carrying two crew members in a pressurized cabin along with instrumentation, food, water, waste disposal and a recovery parachute. The equipment module carried the retrorocket, engines for orientating the spacecraft, and propellant and electrical systems including batteries. It is estimated that the capsule would have made a hard ballistic landing. The Shuguang-1 was a heavy load for the CZ-2 booster, possibly indicating that significant booster upgrades would be required before flight could occur.

Astronauts were selected from pilots of fighter planes based on a combination of political reliability and physical criteria (it was assumed that pilots of fighter planes met the necessary mental criteria). Political reliability, a crucial prerequisite, reflected the political turmoil of the times. The candidate astronauts needed to be less than 1.8 m tall and weigh 80 kg to fit the parameters of the spacecraft. The age was set at under 30 years, but this criterion was flexible for hero pilots (combat veterans) under 35 or 36 years. China all together had fewer than 2,000 fighter plane pilots, and 88 pilots was selected in the first round.

Due to technological constraints, some of the criteria for physical fitness were very demanding, For example, the Chinese astronauts were supposed to experience 10 G in Shuguang-1, thus the overloading test threshold was set at 12 G, which was more demanding than that of the United States and Soviet Union.[13] The selecting process was no doubt an ordeal to the candidates. For example, two candidates' lungs were punctured in the overloading test. Thirty-three candidates passed the physical fitness test and the review committee finally selected nineteen candidates, including an Air Force hero who had shot down an American U-2 reconnaissance plane.

The Air Force was responsible for training these selected candidates, and a specific leadership team, the Astronauts Training Committee (714 Office), was formed on May 15, 1971 for this mission. The candidate training was planned to start in November 1971 and the Shuguang-1 manned spacecraft with two astronauts on board was to be launched by Dong Feng 5 missile in 1973.

Like any other team in charge of a project of national priority, 714 Office was very efficient and all were on the right track. But disaster came. On September 13, 1971 Lin Biao died mysteriously in an air crash in Mongolia, and many Air Force personnel were consequently purged due to their close relationship with Lin. The Astronauts Training Committee was dissolved in mid-November 1971; Qian Xueshen, the program leader, was forced to make self-criticisms due to his expressions of support for Lin before; and the astronaut training program was aborted. The taikonauts were returned to their military units and the 714 program was completely shut down on May 13, 1972.

The secrecy surrounding this aborted program was equivalent to that surrounding the Soviets and the question of whether they raced to the moon or not, which was never clear until after the collapse of the Soviet Union in 1991 and their archives were opened. Even the eighty-eight pilots selected in the first round did not know their mission at first; they were kept in an enclosed building, and communication with their families was prohibited. The West was not made aware of the 714 program until the 1990s. The Soviets were aware enough that they changed their early space station name from Zarya (dawn in Russian) to Salyut so as not to irritate the Chinese. For the Russians, Zarya as a name did not fly until they joined the ISS program in the 1990s.

Despite the apparent demise of the manned program in 1972, work continued, albeit slowly, on the technology for a recoverable satellite (FHW), the necessary precursor to crewed flight. The first FHW flew on November 26, 1975, behind schedule by four years but finally launched. This was the critical technology that had to be mastered if a manned space program was to come into existence. In one sense, this signaled that the Chinese, with a lot of work and money, could in fact achieve human spaceflight. The Shuguang-1 was to be an upgraded variant of the FSW satellites. In February 1978, China admitted it was working on a manned space program. This was reported as the opening stage of a manned program which would eventually grow to encompass a "Skylab"-scale space station.[14]

In December 1981, the head of the New China Space Research Society announced that the crewed flight program was being delayed for reasons of cost. This reflected changes in the Chinese leadership with their focus on the Four Modernizations (discussed in Chapter 4) – the building of China's scientific and technical infrastructure became the overriding priority, and applications such as space technologies were deferred. Apparently, the Chinese human spaceflight effort went into hibernation although activity continued in the background. The desire among the technologists to engage in crewed spaceflight activities remained strong but the economics were politically too difficult. Grondine reports that periodically statements about the possibility of human spaceflight would appear in the press but nothing substantive happened because of the cost parameters.

Project 921 – the second program

In 1989, the first signs of real change appeared with the opening of a feasibility study under the auspices of Project 863, the nation's major technology development program. Aerospace technology was identified as one of Project 863's seven major areas of emphasis, Project 863-2.[15] Within the Project 863-2 rubric, two distinct emphases were identified: Project 863-204 involved development of the launch vehicle and manned spacecraft while Project 863-205 envisioned a space station crewed by taikonauts.[16]

Six projects were presented for consideration. The options ranged from a capsule similar to the Russian Soyuz, several variants on mini-shuttles (with or without engines, similar to the US shuttle and the Soviet Buran shuttle), a fully reusable space plane (similar to the early conceptions of the US space shuttle), an aerospace plane, and a Hermes-derived shuttle. The aerospace plane concept employed air breathing engines in order to reach hypersonic speeds before launching a small orbiter. The Hermes-derived shuttle would have employed technologies from the then active European Hermes shuttle program. The Europeans terminated their project in November 1992 which ended that possibility but China had already set off in different directions.

By 1991, discussions focused on a capsule arrangement, bypassing the shuttle concept at this point. The Chinese Academy of Space Technology considered three capsule designs. The choices ranged from a three-module configuration similar to the Soyuz, a variant on the three modules with different arrangements internally, and a two-module design leaving out the orbital or third module. The decision was to develop the first configuration with the orbital module capable of operating independently in orbit for 180 days. This maximized the benefit from the launch since the orbital module could carry experiments or observation equipment for continued use after the taikonauts' return to Earth.

All space activities had to meet the requirement of positive economic outcomes in terms of adding to China's capabilities and economic growth. Space activities in and of themselves, outside the military realm, had to meet this tough standard of being a definite economic benefit given the high cost. How those benefits were calculated in fact changed over time, from the strictly utilitarian calculus previously to one incorporating tangible and intangible benefits while costs remained largely economic in nature. In that sense, China's calculations became more similar to those driving the early US and Soviet space programs.

Intangible benefits are more fluid, including prestige and international standing as part of the purported benefits. At a point in the early 1970s, in the United States, and far later for the Soviets/Russians, the intangibles began weighing less heavily. That can be a moment of crisis for a human spaceflight program; then arguments arise based more heavily on the identifiable economic benefits derived from the program, besides continued employment for scientists and engineers. For China, among the benefits are the pressures exerted to continue upgrading its technological capabilities. Human spaceflight demands world-class technologies, or a government can be severely and very publicly embarrassed.

In 1990, the first public suggestions of renewed Chinese crewed flight activity were published regarding a four-person spacecraft, while more formally in 1992 at a meeting of the International Astronautical Federation (which China had joined in 1980), a new launch vehicle was described along with Chinese plans for a crewed capsule. In that same year, 1992, the State Science and Technology Commission (the highest political level for S&T) announced plans for a manned spacecraft by 2000 with a space station to follow. Project 921, approved on September 21, 1992, became the road map for China's future regarding human spaceflight.[17]

The original goals for Project 921 were:

- The first crewed launch by 2002
- Orbiting of a man-tended space lab in 2007
- Establishment of a permanent space station after 2010
- Launch of the first test spacecraft in 1998 at the earliest
- Three test spacecraft flights prior to the first launch with crew on board

As announced in 1992, Project 921 was originally built around a new family of larger rocket boosters to be built modularly. That is, the booster stages could be stacked and clustered in different combinations, depending on the objective. A total of six rocket variations were proposed initially but only two were actually developed: the CZ-2E and the CZ-2F as the human-rated launch vehicle (see Appendix A). The latter became the launch vehicle for the manned space program, known as Shen Jian or Divine Arrow. The other component of Project 921-1 involved construction of the manned spacecraft to ride on the booster.

Design of the spacecraft changed somewhat when, on March 25, 1995, Russia and China expanded their space ties with a new treaty supplementing their earlier 1994 accord. Negotiations had begun in 1993 when Russia suddenly found its space technologies a valuable source of hard currency. The treaty provided Chinese access to a Russian RD-120 engine, their Kurs rendezvous system, a docking module, a Soyuz capsule emptied of flight instrumentation and other equipment, and the SOKOL space suit. In addition, two Chinese taikonauts, Wu Jie and Li Qinglong, stayed at Russia's Star City for training as instructors for later taikonauts. The agreement was not a fraternal agreement as allegedly happened between the Soviet Union and China earlier in the 1950s, but rather was understood to be a straightforward commercial transaction. The Chinese had not found the Soviets very fraternal, despite the public rhetoric, so the new relationship was more productive, being based on the self-interest of both parties.

Regarding the question of whether the Shenzhou vehicle was a copy of the Russian Soyuz, the answer is that it is a descendant of that original vehicle but not an exact copy. The Shenzhou is similar in that it consists of three components: a forward orbital module, a reentry capsule, and a rear service module. The Chinese were recycling a spacecraft design that worked, a significant saving of time. The United States by contrast in the early 1970s abandoned the Apollo capsule design for the reusable space shuttle, only to return thirty years later to the capsule design. The orbital module remains in orbit up to 200 days after the departure of the reentry capsule. While in orbit, the module can operate independently since it incorporates its own propulsion, solar power and control systems. Like the Soyuz, in principle, the Shenzhou can serve as a lunar orbiter or transport vehicle to a space station, as the Soyuz does presently for the ISS.

Mark Wade of the *Encyclopedia Astronautica* has published one of the clearest comparisons of the two spacecraft. Table 5.1 compares the Soyuz and the Shenzhou.[18] As can be seen from the table, the Shenzhou is larger and more robust in terms of its operations. The Russians have developed the next-generation Soyuz spacecraft, the TM and TMA versions, to service the ISS, a role the Chinese would like to engage in also.[19] Chinese engineers have pushed the earlier design ahead in part because a shuttle-type operation is more difficult to operate and more fragile. In fact, the United States is returning to the ballistic approach to reentry used by the Russians and the Chinese in their new Crew Exploration Vehicle (CEV). The earlier Apollo program used such a return approach but that was abandoned in the early 1970s in pursuit of the space shuttle which was expected to be fully reusable. In fact, the US still pursues the concept of having a reusable CEV at least for several missions, not completely reusable indefinitely. Its "Apollo on Steroids" approach signals an even larger vehicle than before, three times the size of the Apollo vehicles.[20]

Russia's economic woes in the early 1990s made their entire space inventory available to the highest bidder. The United States was buying

Table 5.1 Comparison of Shenzhou and Soyuz spacecraft

	Soyuz	Shenzhou
Complete spacecraft		
Total mass (kg)	7,250	7,840
Length (m)	7.48	9.25
Diameter (m)	2.72	2.80
Span (m)	10.06	17.00
Service module		
Total mass (kg)	2,950	3,000
Propellant alone (kg)	900	1,000
Length (m)	2.60	2.94
Diameter (m)	2.17	2.50
Diameter base (m)	2.72	2.80
Reentry vehicle		
Total mass (kg)	3,000	3,240
Length (m)	1.90	2.06
Diameter (m)	2.17	2.52
Orbital vehicle		
Total mass (kg)	1,300	1,500
Length (m)	2.98	2.80
Diameter (m)	2.26	2.25

former Soviet space technologies especially rocket engines for its new boosters, the redesigned RD-180 (formerly the RD-170) for the Atlas 5 for example. The world however had changed from the 1950s and now China had built its capabilities to the point where the purchases, while helpful, only moved its existing efforts forward faster. Their absence or nonavailability would simply have delayed but not stopped Chinese progress on the road to achieving independent human spaceflight. While the launch technologies and crew equipment were under development, China also constructed new launch facilities at the Jiuquan Satellite Launch Center, including a vehicle assembly building, a transporter for moving boosters and a tower for servicing on the pad.

In addition a total of four Yuan Wang 4 tracking ships were constructed, a necessity given China's paucity of foreign-based tracking stations. The Beijing Aerospace Command and Control Center along with testing facilities was also built in 1993. China has found the pursuit of human spaceflight challenging, forcing upgrades across the board that otherwise would have been years later in construction without the self-induced pressure to move forward. The space program since its inception as a missile program has been one engine driving China's technologists to drastically upgrade their skills and equipment; human spaceflight is but another step in that continuing process of improvement.

The Shenzhou program

The two taikonauts training in Russia at Star City were important not only for the technical expertise acquired but also as a public symbol of Chinese confidence that crewed spaceflight would in fact happen. China's manned space program was not transparent to outsiders but the fact that more information was being released was one major marker of their growing confidence in their capabilities.

In 1996, two important steps occurred: the first public announcement of Project 921 and the selection of the twelve taikonauts to join the two instructors in training for a future spaceflight. The taikonauts, in a manner similar to early Soviet cosmonauts, remained publicly unknown as individuals, unlike the Americans with their Mercury Seven astronauts who became public figures well before their first flight in 1961. These differences in approach reflect different governing styles but also the reality that China's race to space was only against itself. Unlike the United States, during its Cold War competition, the Chinese government retained control over all information, helping to insure that its manned space activities appeared only in a positive light. The United States desired the same positive publicity but was forced to endure the bad with the good due to the openness of the Kennedy Space Center to surveillance by news media.

The proposed spacecraft, the Shenzhou (or Divine Vessel), represented a real challenge. China had consciously demonstrated its capacity to launch payloads to orbit and retrieve those payloads intact. But now, given the greater publicity associated with human spaceflight, the pressures for success were much more intense. Chinese space program leaders proceeded cautiously, at least in public terms, in the development and testing of the new spacecraft.

Their goal was success; failure was unimaginable given their lofty aspirations. The same considerations that motivated Mao Zedong in the 1960s – international prestige and clear evidence of Chinese strength – continued into the new program but were now built on a firmer technological foundation. Evidence as to their seriousness can be seen in the financial commitment made to the program. One estimate is that China spent approximately $2.1 billion from 1992 to 2003 on the Shenzhou program, with half spent on facilities.[21] By US standards, the amount is small, but for China it was huge. More critically, the total drastically underestimates purchasing power and differentials in prices.

Early issues in the program included the launch escape system – the original system failed to work properly. The problem was not dissimilar to those confronted by the Soviets. The escape system is a rocket motor that fires in the event of a launch pad incident – lifting the crew up and away from the pad area with a parachute deployment for landing. The Soviets found their system temperamental and potentially dangerous to the crew. By October

1998, that problem was addressed with a successful test launch escape. From that point forward, the Chinese were more confident in their ability to protect their crew in the event of a launch problem.

Developing the capsule itself took time since political tolerance for failure was so much less and more unforgiving. This was also true for the Americans; after the Apollo 1 pad fire in January 1967 resulting in the deaths of three astronauts, a purge occurred of those managers in charge. The political pressures were intense, failure was intolerable. In fact, the Shenzhou 1 launched on November 20, 1999 resulted from a change in plans. The CZ-2F booster was ready to fly by 1999 but no spaceworthy capsule existed. In order not to lose program momentum, a test version of the Shenzhou capsule was upgraded to flyable prototype status. The decision was to upgrade the capsule enough for the shell to fly to space and back. The spacecraft carried no instrumentation beyond guidance and recovery equipment – meaning no experiments or test dummies in order to test flight effects were included. For the program, that was a critical test because the recovery aspect of the flight has always been the most difficult and hazardous.

The Shenzhou 1 flew for fourteen orbits in its simple no-frills flight with no maneuvers during flight. The major problem was a failure to transmit a command to the onboard computer for initiating reentry. On the final orbit, success was achieved, with an acceptable landing occurring on November 21, 1999. The Soviets had gone through similar problems in their early cosmonaut flights when reentry became problematic.[22]

The success of the first flight, however, did not lead immediately to a second flight – unlike the early space race from 1958 until 1969, there existed no external competitive pressures. China had to get it right rather than fast and right, unlike the Soviets and Americans. This difference in pace also reflected the Chinese concern with costs, unlike the perceived life or death competition operating during the Cold War.

The flight of the Shenzhou 2 was in fact the first of the three scheduled test flights prior to the first actual manned mission. The entire capsule was more functional including an operational orbital module. Launch occurred on January 19, 2001 after delays caused by possible damage to the CZ-2F booster. The spacecraft flew for seven days or 108 orbits in what was labeled a "standard Shenzhou orbit" which involves maneuvering in orbit in order to create a circular orbit, this time at 330 km.

The Shenzhou 2 flight, however, contained a possible problem, not publicly acknowledged at the time, which was that the reentry capsule was never seen after its return – no photos were released of the returned vehicle – implying some damage had occurred. It is assumed that the parachute landing system did not completely deploy, with a hard landing the result. After Shenzhou 2, China was publicly committed to its human spaceflight effort – failure to follow through would have publicly devalued its past space accomplishments.

The Shenzhou 2 reentry question reinforced their caution but meant the Chinese had to push forward in a reverse prestige sense. This was similar to the situation or the Americans who were publicly embarrassed by their first satellite launch attempt, the Vanguard launch failure in December 1957.[23] The Vanguard flew about four inches off the launch pad before falling back in flames and explosions, a debacle televised live before the entire world. China's problem with reentry was more private but the potential for humiliation was real, and it increases even more as the stakes rise. Reentry remains a difficult and dangerous task which can result in injury or death to the crew. The Soviets several times had accidents and hard landings where cosmonauts died. The United States most famously lost the shuttle Columbia during reentry and lost a Mercury capsule on an early flight when it flooded after a hatch was accidentally opened. The orbital module, once separation occurred from the reentry capsule, moved to a higher orbit in order to conduct various orbital maneuvers before entering the atmosphere on August 21, 2001.

Unlike the first mission, Shenzhou 2 carried sixty-four experiments, both in the orbital module and in the reentry capsule, with additional experimental packages attached to the orbital module's outer shell. China cannot afford to "waste" a launch to orbit, so sending multiple experiments to orbit in addition to the human side of the flight becomes its normal practice, similar to the early US and Soviet programs. Experiments were drawn from the disciplines of materials, life sciences, astrophysics, space environment physics, and other microgravity sciences.[24]

The Shenzhou 3 flew to orbit on March 25, 2002 as the first crew-rated spacecraft, albeit with no crew. That meant that all the systems necessary for a human to fly in the vehicle were in place. A working escape tower system was in place for the dummy passenger who flew to orbit. The flight however was delayed when an electrical connector on the CZ-2F failed, necessitating a rollback of the launch vehicle to the Vehicle Assembly Building. The mission incorporated scientific payloads covering a wide spectrum of areas. "Shenzhou 3 carried 44 science and utility payloads into space. The science experiments involved material and life sciences studies, Earth and atmospheric observations, and space environment monitoring."[25]

The next mission, Shenzhou 4, was the final rehearsal for the first manned flight to follow. Lifting off on December 30, 2002, the escape tower was active, ready for use in the event of a launch incident, similar to the next crewed flight. The flight was routine in the sense that no noticeable or reported problems arose, with the capsule return occurring on January 5, 2003. "Shenzhou 4 carries an assortment of biology, biomedicine and physics experiments. In addition, the vehicle is outfitted with a multi-mode microwave experiment for remote-sensing survey of the Earth and the space environment."[26] Reflecting its growing confidence in the human spaceflight program, China announced that the first manned flight, Shenzhou 5, would occur within a year. This clearly put the pressure squarely on the program to deliver success with the next flight.

On October 15, 2003, the Shenzhou 5 (Divine Vessel 5) flew to orbit with the taikonaut Yang Liwei on board, the first act in China's human spaceflight program. His orbital flight lasted over twenty-one hours or fourteen orbits. The launch was widely anticipated, with the Chinese government undecided until near launch time whether to provide live television coverage or not. The possibility of failure triggered the usual government reflex of secrecy. Their decision was not to do so but video footage was released immediately after liftoff. The importance of the mission was further highlighted by the fact that no scientific experiments flew on it, precluding any possible accidents caused by extraneous factors. In fact, Yang basically sat in his seat for the entire flight, minimizing the possibility of mishap. The orbital module carried two cameras for imaging the Earth and remained in orbit until May 30, 2004. The inclusion of the cameras reinforced some views that the Chinese military was using the Shenzhou program to advance military purposes. That is obviously true but is considered no different from the early Soviet and American space efforts.

With completion of this first manned flight, China exploded with nationalistic fervor. The government interpreted the flight as further ratification of their status as a global player – a challenger to the United States given their shared history of mutual suspicion and antagonism. Within Asia itself, China was widely perceived as having supplanted Japan as the dominant political player. The perceived political benefits outweighed any immediate concerns about cost.

Joan Johnson-Freese interpreted the flight as having a great positive impact upon China's image in the world, with a concomitant decline in US prestige.[27] Her argument in part was that China's success occurred at a time when the US space shuttle was grounded due to the Columbia accident. From this perspective, the United States lives on its past glories surrounding the Apollo program, but for well over half of the world's population that success is ancient history, occurring well before they were born. The United States is perceived as falling behind China especially as the shuttle return to flight hiatus lagged for nearly two years (in fact requiring another year for the second mission to validate the first) plus the shuttle itself by presidential directive goes out of service in 2010. By contrast, China's young space program appears to be on the rise.

Some contend that China's success will in time kindle another space race, the US and China being the protagonists with other states more likely observers and followers. The US Vision for Space Exploration announced by President George W. Bush in January 2004 is seen as the bugle call for the new race.[28] Given the Vision's budget issues, China's program appears unlikely to kindle the next space race. In fact, other states are being invited to join, including possibly China, as will be discussed in Chapter 6. The more likely space race will concentrate on military space activities since both states, especially their militaries, find such activities essential for their military success.

Now what? – the short term

China's successful launch and return of taikonaut Yang Liwei in October 2003 immediately raised the question of what to do next. Ironically, their public success raised questions of whether China's goal became pushing forward aggressively to incorporate new activities, or continuing to move slowly, incrementally building up to the next level. Deciding this question was crucial since the wrong choices could dissipate scarce resources, always an important issue for China, without actually accomplishing much. As discussed above, some signs of euphoria appeared in the immediate aftermath of the Shenzhou 5 success, when an avalanche of new programs were proposed as if they were all going to be accomplished immediately (see Chapter 6).

The immediate first priority ironically for China becomes simply doing it again – sending another crew to orbit and safely returning them to Earth. Further illustrating the difference between China's human spaceflight effort and those of the two space pioneers is the fact that the second manned flight was put off for two years, until October 12, 2005. The Soviets and Americans truly competed in a space race, the devil take the hindmost. For China, there exists no immediate space race challenger while the United States and Russia are immersed in their own space programs – the Americans are dealing with return to flight issues, while the Russians seek to stabilize funding for continuing their efforts. China, through the Shenzhou 5 success, has reached its immediate political objectives in terms of prestige and heightened perceptions of China's strength. Therefore, there is no rush for an encore, rather a measured pace can be pursued, building sequentially upon the last building block.

Some have argued that repeated US problems with the space shuttle have affected Chinese perceptions of how fast they should move. The argument is that NASA is in transition from the shuttle to the CEV, with long-term program stability unclear. This opens the door for China to shine with its obviously successful program. For much of the world, the Apollo program is very old news while China presents an image of vitality and strength. Given the success of the first flight, followed by a successful Shenzhou 6 mission, the temptation to accelerate their program may prove overwhelming.[29]

The Shenzhou 6 mission, launched on October 12, 2005, from China's perspective placed the stamp of success on its efforts. Two taikonauts, Fei Junlong and Nie Haisheng, flew on a five-day mission. During that mission, they moved about the spacecraft and conducted several experiments.

This constituted the second act of China's manned space program – repeating the same task with a greater degree of difficulty.

One of the questions arising in the runup to the mission was whether Yang Liwei would be on the mission. One demonstration of China's confidence in their flight technology will be seen in whether the first taikonaut Yang will be allowed on successor missions. Yuri Gagarin, the first man in space, never flew to orbit again, only to perish in a flying accident. John Glenn, the first American to orbit the Earth, was grounded by presidential order from flying again, causing Glenn to leave NASA and embark on a political career in the US Senate. Glenn later flew on a shuttle mission in his seventies as a test of older people in space. Whether China exercises similar caution will be an early informal test of its confidence. Having your pioneer die on a subsequent flight is a terrible public blow to a new program.

The Shenzhou 6 mission's return to earth left its orbital module in space, operating successfully for well over a hundred days. The spacecraft continued experiments plus conducted photographic missions – again reviving talk of its dual-use purposes. China had now experienced success on all dimensions in terms of orbital operations, leaving only the Shenzhou 7 mission as the windup for this phase. What China does beyond Shenzhou 7 becomes somewhat of a puzzle due to Chinese secrecy, although the general pattern appears to be falling into place.

The Shenzhou 7 mission now appears clear; it will carry up to three taikonauts and a space walk or extravehicular activity (EVA) will occur during the mission. The expected date of the mission, however, has been pushed back from October 2007 to some time in mid-2008. This delay appears related to the 2008 Beijing Olympics when maximum public attention will be focused on China. Prestige and goodwill are expected when greetings are sent from Shenzhou 7 to the participants, as happened in Athens in 2004 from the ISS. The length of the mission is expected to be between five and seven days. How many EVAs will occur is unclear although it appears one taikonaut will space-walk for an hour or less, since the purpose is testing the Chinese version of a Russian space-suit design.

The Shenzhou 7 will carry an operational androgynous docking port mechanism which will allow mating of two spacecraft. Any docking at first would likely involve rendezvous with the previous spacecraft's orbital module left behind during reentry, although the Chinese could mirror other states' earlier flight experiences by tracking and docking with target vehicles launched separately. This mission will be a precursor to China's efforts at a small space station or, more realistically, space habitat. Given the success of the first two missions, and assuming the success of Shenzhou 7, this places China in an interesting position relative to future directions.

Since the Chinese are employing the Russian docking design this in principle allows their Shenzhou spacecraft to dock with the International Space

Station (ISS). With the pending shutdown of the space shuttle in 2010, the Chinese may have some opportunity to help service the ISS – expanding the possible vehicles flying to the ISS beyond the Soyuz and the European Automated Transport Vehicle (ATV). The United States is accelerating development of its Crew Exploration Vehicle (CEV) so that it does not lose the capacity to reach orbit. The original time configuration was for a four-year gap between shuttle shutdown (2010) and the CEV becoming operational (2014), leaving the Soyuz the only current service vehicle for the ISS.

China becomes the wild card in that the Shenzhou theoretically provides another option – one that works while the European Crew Return Vehicle is still in development. The political backwash from this may be substantial since the United States has long resisted efforts to incorporate China into the ISS, the ultimate symbol of international space cooperation. However, American clout will decline drastically when it effectively abandons the ISS in pursuit of its next dream.

For an advanced space participant, and China has become one, the missions beyond Shenzhou 7 become the key to defining future directions. China has, at least since the second manned program was initiated, spoken repeatedly in terms of a space station. This is the third act in human spaceflight. As they decide how to do this, the Chinese encounter two pathways to the future. The Soviets and the Americans both started by launching small space stations to orbit, usually in one payload, The American Skylab was an aberration growing out of NASA's desperation to keep its manned space program visible to the US Congress and public. That effort saw three crews live on the Skylab before the United States lost its ability to access the station, which fell from orbit in 1979. Not until the approval of the ISS in 1984 did the United States build another space station.

The Soviets/Russians inaugurated a series of small space stations both military and civil at first, the Almaz–Salyut series.[30] The military aspect, the Almaz, faded into the Salyut program as the severe limitations of space stations for military purposes became clear. Their program was much more systematic and productive than the American effort. In 1986, the Mir space station was built with additional modules arriving at Mir up until 1995 when a docking module was delivered in order that space shuttles could visit the station. However, the Mir 2 replacement station was never built, with Russian economic problems putting a stop to the project. The core Mir 2 component was later recycled for use in the ISS.

The ISS began as a single-state space station (three international partners were added later), even being named the "Space Station Freedom" in 1988, but persistent program failures led to the program's internationalization beyond the original three international partners, ESA, Japan and Canada. Russia was added to the mix in 1994 in order to access their space hardware, plus tie the Russian Federation with its new democracy closer to the West.[31] The ISS program has expanded its international roster to around sixteen states. China has indicated interest in joining the program but the US reluctance to incorporate China has kept it on its solitary path.

The Chinese space station apparently was earlier envisioned as a large-scale space station – the equivalent to the completed ISS or Mir in terms of size and capabilities. That option has real costs associated with it plus much greater technical risks. The construction of the ISS illustrates the costs and risks. A space program such as China's with its tighter budget and greater concerns about public image is one that finds the large-scale space station an unacceptable risk at this time. Therefore, the option being pursued by China is a variation on the earlier experiences of the Soviets/Russians.

The space station concept presently (although still subject to change) involves using Shenzhou spacecraft in order to construct a smaller modular space station or habitat. As presently configured in more recent public statements, the Shenzhou 8 mission scheduled for 2010 will be a space lab module with two docking ports (with a gross weight of about 8 tonnes). This weight and size can be lifted using current launch vehicles or slightly upgraded versions of the CZ-2F (or CZ-2G). The Shenzhou 8 becomes the building block for the space station complex.

The present plan is that the Shenzhou 9 will be launched within a month of the Shenzhou 8 without a crew. The Shenzhou 9 will dock with the Shenzhou 8. Subsequently, the Shenzhou 10 will launch with a crew, creating a man-tended space station. Man-tended refers to a structure visited and worked on by a crew who do not permanently occupy the site. This is what would have happened to the ISS after the 2003 Columbia accident if manning of the station by only two personnel had not been feasible. Originally, the understanding was that a minimum of three personnel were required to maintain the ISS. The ISS would have been visited on occasion until the shuttle returned to flight, rather than being permanently manned as presently occurs (spring 2006). This approach provides a lifeboat for the crew in that the Shenzhou 9 is available if a problem arises with the Shenzhou 10. Alternatively, one could begin sending successive crews up in the replacement vehicle while using the earlier Shenzhou docked to the Shenzhou 8 for transportation home, similar to the ISS rotation of Soyuz spacecraft. This approach would assume a much higher flight rate than currently projected but in principle is possible. It also makes sense in terms of maximizing the usefulness of the space station. One item not publicly discussed is the Chinese equivalent to the Russian Progress resupply spacecraft.

Other space station options have been suggested that are equivalent to the US Manned Orbital Lab (MOL) developed for the US Air Force in the 1960s and the Soviet TKS system. Neither actually flew operationally. In the former case, the reconnaissance surveillance aspect was superseded by robotic spacecraft transmitting images from space. Those programs both envisioned the destruction of the small space stations at some point during reentry – not an economically sound choice for a program whose budget is closely monitored by the central leadership.

Regardless of when a space station actually flies to space, China will clearly move up even farther prestigewise. The world has witnessed twenty plus years of US frustration and lack of progress on its space station program. The Chinese version will likely not be equivalent to Mir or the ISS but, in the prestige wars, that fact will not be relevant. The question is program momentum and the image of continuing progress and expanding capabilities.

Concurrent with the Shenzhou program, China has also embarked on a program aimed at accessing the moon.[32] The Chang'e program has started the construction of its first lunar orbiter, Chang'e-1, with launch expected in April 2007. The program is aimed at reaching the lunar surface and returning a sample to Earth. The projected time frame appears to be 2014 to 2017 for the sample return mission, assuming no glitches arise and funding is available.[33] The Soviets successfully conducted three sample return missions, while the United States sent astronauts to the lunar surface who returned samples. The program is critical because it is the precursor to a Chinese manned mission to lunar orbit, and eventually the lunar surface, which is even more distant – 2030 or beyond. The program also draws some caution from Chinese experts for pragmatic reasons of limited resources and priorities if China is to continue operating a space program of value to the society. These conflicts are reminiscent of similar disputes within the American space program over priorities.[34]

This effort also occurs at the same time as the United States, ESA, Japan and India are all engaged in lunar missions as the forerunners to a return to the Moon. This latter effort is keyed in part to the long-range goal of humans reaching Mars, the US Vision for Space Exploration (VSE). Estimates of how fast the VSE program will move down this path are still fluid. Some argue that a manned trip around the Moon could be accomplished by 2012 at the earliest but more likely 2014 or later. The actual landing and return appears to be in the neighborhood of 2020 or later, barring unusual circumstances. China's lunar effort presently is independent of this larger effort but, by its activity, China draws attention. This American exploration effort comes at a time when China has the potential to participate as a significant partner.

What is important is that China now openly speaks of its ambition to compete at the highest level and evaluates its chances as excellent to eventually duplicate or surpass its putative rival, the United States. China by necessity and increasingly by choice chooses to be a solo operator, although its links in other areas of space activity (i.e. Galileo) may create more opportunities to cooperate. Cooperation for China, however, will now come as an equal partner or not at all.

Conclusion

What is fascinating, thus far, is China's apparent ability most recently to keep its aspirations and ambitions in alignment with its politically available resources. As was discussed above and will be addressed again in Chapter 6, China's future plans are extensive but the budget pressures force real choices – the choice presently is to proceed forward but at a measured pace. The Chinese are becoming engaged in the state's space program, which accentuates pressures to assertively push forward but concurrently increases the pressures to move too quickly. China may not be in a direct race with the space pioneers but it is in a race with its own ambitions.

Chapter 6
Assessing China's future in space

Introduction

China's quest to enter outer space has obviously been successful. Since April 24, 1970 onward, China has at some level been a space participant although its activity levels varied widely over the years. There was no linear progression of success but rather a zigzag pattern in pursuit of their goal, and the goal itself morphed over time. More recently, Chinese activities have grown progressively more sophisticated and more robust. Now, China has become an increasingly active and consistent space participant. In fact, its earliest activities at first drew limited public attention, in large part because of China's secrecy and international political isolation. Some attention was paid to Chinese space activities by the superpowers and China's geographic neighbors but their focus was primarily on the potential military aspects (e.g. missiles) rather than any peaceful space activities.

For the US government, that intense concern with the military aspect continues to the present even though the American private sector in time came to see Chinese launchers as viable alternatives and competitors to the established launch vehicles. The launch failures in the mid-1990s damaged China's standing as a launch provider but recovery is underway. Continued launch success, now over forty consecutive launches, will help alleviate insurer fears. The export license imbroglio, however, has thrown that aspect off-track since US satellite vendors constitute a major segment of the market. Any loosening of American regulations will take time. Reports such as the Pentagon's 2005 report to Congress regarding the PRC's military power will further limit any fluidity in the situation. The report indicated that China's military power is growing and may be used to advance its objectives through use of force or coercion.[1] The 2006 DOD analysis is even more skeptical of Chinese intentions especially with regard to space activities.[2] Politically speaking, such hardline analyses freeze the present situation in place.

China is not waiting, given its pressing economic and social priorities. As a result, any Chinese success in the commercial market will now be driven by its perceived reliability as a launch provider (the insurance issue); other areas of space commerce are still under construction. The Americans remain the major international exception, with domestic political considerations driving their receptivity to Chinese commercial offerings. It is against this global background that one must assess China's future and its impact on the system.

Public space activities almost by definition are built on some national vision of the future – usually one of technological excellence accompanied by accelerated national economic growth. That vision must be strong enough to justify the heavy sacrifices demanded, especially for a developing state. The economic plenty that is expected to follow as a matter of course is usually the carrot driving the Chinese space program. China's space program presents a riveting example of an underdeveloped state executing such a futuristic vision and what must be overcome in terms of obstacles and sacrifices. What is particularly important is the long-term nature of the commitment and the sacrifice required for success. The hyper-frenzy of the first space age – the superpower space race – often blinds us to what is truly required in terms of resources. Most developing states, such as Brazil, are unwilling or unable to make the necessary commitment, so their success comes more slowly.

One must acknowledge that the development of Chinese space activities only in retrospect appears to track an inevitable successful outcome. As was described earlier in Chapters 3 and 4, China's efforts proved erratic at times even though some degree of political support persisted, even in the depths of the chaos of the Cultural Revolution. In the early days, that political support was based more on the military value of space launch – missiles – rather than on the exploration or exploitation of space for national development. Zhou Enlai, the great survivor in the Chinese political arena after 1949, was too often crowned with all kinds of honors, while political losers, such as Lin Biao, were divested of any credit. This game of saint or devil was ironically further reinforced as rational choice even in the circle of scientists in China. In order to survive in terms of receiving funds and achieving desirable social status, Chinese scientists vigorously seek their political patrons alive or dead physically but alive politically. Thus it is not surprising that Zhou Enlai's contribution is widely known, while that of Lin Biao's faction was secret for many years until very recently when Wang Bingzhang's name was again seen in the media. Interestingly, Deng Xiaoping's contribution was not that he favored the space program; there is no evidence that he was a big fan of space technology, but his leadership created the requisite stability that made success possible.

China's size (both geographically and in terms of population) makes it unique; only India truly compares, but the economic struggles China has endured to this point make it more truly representative of the many states now emerging into the global political and economic system as significant players. The two superpowers and the rest of the developed world still operate in a very different environment. In a fundamental sense, China represents one model for those other states both negatively and positively. This reality must be understood in analyzing the future development of international space policy. The bipolar world of the past, which drove the development of the original space legal regime, is gone but its effects linger on into the present. Therefore, China's future development of its space program will, as has occurred in the past, remain a work in progress.

To stop is to stagnate, leaving Chinese society disarmed militarily and economically among its competitors. The myriad possibilities for development are greater than the available resources so for China the question becomes how to balance such competing pressures. Those pressures can be seen in the recent approval of China's fifteen-year S&T plan in which manned moon exploration is a major priority. This is part of China's effort at revitalizing its entire research establishment including the Chinese Academy of Sciences.[3]

In this chapter, we will highlight, first, China's impact on the international space regime; second, China's future policy directions will be discussed in light of its stated aspirations; and third, we will identify possible pitfalls embedded in that future. The Chinese space program remains a work in progress and plans will change; the intent here is to point out issues that are coming into focus or will do so within the next decade.

China's impact on the international space regime

The international space regime struggled into existence after the first satellites were lifted into orbit, starting in 1957. The field during its construction was dominated by the two space superpowers, the Soviet Union and the United States. Other states such as Canada and the Europeans were heavily involved in setting up the rules of the road, such as they were, for space activities, even though the effective limits became whatever the superpowers were collectively willing to accept. Without their support, little could be accomplished.[4]

The result was a legal regime, which effectively ratified what was possible rather than what was most desired by states other than the two superpowers. Politically, the other states' collective problem was that separately or collectively none of them could really compete with the first two during that time period. What was clear but hidden at the time was the unique political conditions that drove the situation: a two-power global nuclear arms race. That perilous situation was dramatized by the Cuban Missile Crisis in October 1962 when an unforeseen nuclear confrontation drove the two powers to the brink of nuclear war. Both in effect stepped back from the abyss. Once that nuclear edge was finally dulled, the ancillary two-power space race's intensity quickly dropped. The military nuclear arms race continued, and civil government space activities continued, but at a slower pace.

In fact, the space race of the 1960s was never critical in terms of military capabilities; rather its contributions were primarily psychological and prestige-driven. But, by the end of that race, the fundamental ground rules for conducting any and all space activities were put in place. China was well outside that circle until the 1980s and 1990s. Ironically, the fact that China was excluded did not disadvantage it in the long term. The rules forbidding weaponization of space meant that the possibility was forestalled until now, when China is capable of participating.[5] For reasons of their own strategic interest, the two space powers rejected space weaponization, putting off consideration of the question until later. Since then China has become a full participant in the debate and capable of participating at some level if such weaponization occurs.

The Europeans collectively or separately were unwilling to compete at first, even though they bridled under American restrictions.[6] The European Space Agency became their organizational vehicle for acquiring a certain level of technological independence, and their recent efforts have proven very productive. Japan's space efforts initially in the 1950s were constrained by various domestic and international restrictions growing out of their defeat in World War II. More recently, Japan's space activities have been constrained by domestic budget woes and technology failures. Space activities still remain subject to the frailties of space technologies – the operational margins remain thin as China has discovered repeatedly although with decreasing frequency more recently. China's growing skill in executing its space activities has immeasurably raised its international prestige and influence.

China's arrival on the international space scene was a long anticipated event, in part because its earlier political isolation and policies initially kept its intentions so mysterious. That mystery fostered attribution of great hostility to China if ever released from its isolation. China reinforced that rogue image during the Cultural Revolution when the society appeared to self-destruct. Rhetorically speaking, China projects itself as the oppressed state struggling to overcome the hostile actions of the United States specifically, but also with regard to other developed states who perceive China as a political rival or potential economic competitor. Obviously, the United States and China cooperate in terms of trade but tensions exist, especially when the United States is perceived as damaging China's space endeavors, as in the export license controversy.

In terms of its immediate impact, China changes the political climate within which space activities operate in two ways. First, China's successes in achieving Earth orbit in 1970 and later independently lifting a Chinese national to Earth orbit in 2003 raise the aspiration levels for other economically challenged states contemplating space activities. This assertion was made earlier in Chapters 1 and 2 but now, having described the extreme difficulties encountered during China's journey to outer space, one can appreciate more fully what such achievements entail. States such as India have clearly set off on a similar path for the same mix of international and domestic reasons, but China's example of go-it-alone development has particular resonance even for India, especially when that state encountered external restrictions such as the MTCR which temporarily derailed its space efforts. India had to develop indigenous alternatives to the Russian technologies it had attempted to purchase but had been denied.

For other states even more technologically and economically challenged, China provides an example of how a single state can systematically overcome many of those deficiencies. The hierarchy discussed briefly in Chapter 2 is relevant here as states lower down in the hierarchy strive to improve their space capabilities. What China offers is both an example and a role as a facilitator for those other states. Its launch vehicles are distinctly less expensive than other states' launchers. In addition, China's interest in cooperative space activities makes it a likely partner for such states, although that decision remains a political judgment, not entirely or necessarily a scientific or technological one. By partnering with China, costs are lowered for these less advanced states. Currently, China places fewer political restrictions upon its partners than does the United States. China has arrived at the political position where its space activities are still employed as political devices but the political purposes have changed from defensive to proactive. That is, space activities become one mechanism by which China now attracts allies rather than simply fending off its purported and real enemies.

Within the developing world itself, China continues as a space leader with much greater perceived prestige and power than previously. For example, China has partnered Nigeria on several projects, the most recent being its building of a comsat for use by Nigeria; likewise for Venezuela, another oil-producing state. The importance here is twofold: first, Nigeria is an oil exporter whose oil is sought by China; and, second, the Nigerian comsat is China's first sale of a domestically produced comsat to an international customer.[7] What is unknown is whether, in time, China will acquire some of the characteristics of a "have" state compared to other less technologically sophisticated states: that is, less sensitivity to others' pride or needs.

For developed states, the Chinese effect was initially less self-evident because those states perceived themselves as being among the advanced states. Their perceptions are being restructured because China, especially for the Europeans but also for others such as Brazil, has become an attractive alternative to the United States with its erratic and arbitrary (to outsiders) political-military restrictions. Regarding China itself, it has crossed the threshold of being considered just another aspiring space participant to a position of equality with many established space participants. This fluidity reflects the openness of the field of space applications. With the requisite resources (a big caveat), any one can play.

This new status opens up possibilities for China that did not even exist a decade or so ago. One prime example is China's active participation in the European Galileo navigation satellite system. China's financial contribution is 200 million euros, which makes it a major player at the startup or developmental stage, rather than just another customer or latecomer to the project.[8] Most economically challenged states are normally valued only as customers rather than active participants in a particular system's development and maturation. Other states such as Israel, South Korea and India have also become engaged in the program.[9] There are obvious military implications embedded in the Galileo project but the commercial aspects are paramount for the Europeans, at least in the short term. China's participation clearly signals its increasing acceptance as a member of the space elite. This permits access to cutting edge technologies, something which China still needs if it is to continue improving its space applications.

China started far behind relative to the two space pioneers – enormous energy was expended closing that gap, at least selectively. Moving up from its status of selective equality with the space leaders demands another level of heightened effort – one that can be best leveraged through cooperative programs with other states possibly stronger in certain technologies than China. Through the synergism of such alliances, China can more quickly improve those areas where it is weaker. This strategy is not just one for lesser states but occurs at all levels in the sphere of space activity. The United States, for example, is no longer the dominant technological player in many areas of space activity. For the Americans, this situation represents a crisis but one unlikely to be reversed in the short term given escalating budget deficits.

Second, China's newly prominent political and technological capabilities improve its ability to articulate its policy positions relative to the existing space legal regime which China perceives as advantaging the presently dominant space power, the United States, and as a consequence limiting the efforts of others. The disputes cover the range of space policy but, in the short term, China's focus is on military space applications. China's goal is to delay if not prevent significant changes in the regime, especially those understandings and formal agreements rejecting actual placement of weapons in outer space – the weaponization of outer space.[10]

This admittedly self-interested position places China on the side of important states such as the Europeans who are opposed to persistent American advocacy of the weaponization of outer space. The space age from its beginnings contained the possibilities for such weaponization but the United States and the Soviet Union for their own separate reasons accepted the concept of a "sanctuary" in outer space: a zone within which no weapons are permitted including weapons of mass destruction, especially nuclear.[11]

The sanctuary concept became embedded in the space legal regime along with the treaty-based understanding (the Liability Convention) that states are responsible for whatever their nationals (vehicles or individuals) do while operating in outer space.[12] That stance translates into state control over all space launches (whether commercial or public) and their payloads. For all states, there existed no truly independent commercial launch industry so control was comparatively simple. The 1960s and 1970s perspectives emphasized state control for national security reasons and opposing political ideological perspectives. The Western states argued for control for security reasons while the Marxist states advocated control for reasons of ideology and security. Initially, that control reflected governments' monopoly over launch vehicles but the Liability Convention effectively eliminated purely private space activities. Any private space activities operate presently only on the sufferance of the government which must authorize their launches and other activities. Space activities remain fundamentally outside the market economy approach that is reshaping the global economy. This reality means that space markets are at times distorted by noneconomic agendas. For China, the short-term effect is to hamper development of its space industry by allowing other states to, for example, subsidize their launch fleets and place export restrictions on space technologies. The overall impact slows China's ability to grow economically, utilizing their space technologies.

Militarily speaking, the sanctuary concept was originally interpreted as not excluding military satellites, just spacecraft carrying weapons. This interpretation, however, is not embedded in treaty form – only weapons of mass destruction, specifically nuclear, are specified. Past practice has developed into an understanding or norm that no weapons are allowed in outer space.[13] The Soviets and the Americans placed military satellites in orbit but none carried weapons; in time and through repetition this became the *de facto* standard. No other state was involved initially so there was never a politically viable alternative interpretation to their standard: military satellites – yes, weapons – no. That view permitted reconnaissance, early warning, navigation and communication satellites to be placed in orbit in order to enhance the effectiveness and lethality of military forces operating in the atmosphere. In principle, one could have two states fighting each other while their respective satellites operate side by side in orbit. No military conflict has yet engaged two or more states in opposition to each other which all possessed significant space assets. The 1991 Gulf War, touted as the first "space war," was only a space war on one side, that of the Coalition, and not of Iraq.[14]

For China, this discussion of weaponization is not a trivial one since modernization of its military remains an expensive and difficult process. Chinese support for the sanctuary concept becomes from their perspective an important tool for keeping the strategic balance more even than otherwise possible relative to the United States. Chinese space activities add further credibility to their protests because the implications are that China in the future can engage in similar weaponization activities. Early Chinese statements on the issue were irrelevant given its lack of credibility as a space participant. The Europeans are discovering that their credibility *vis-à-vis* the Americans improves as they acquire more military space chips to counter US programs. The Galileo navigation satellite system for both Europe and China is in fact a declaration of further political, economic and military independence from US influence or control. China is not just another voice but one whose potential military power is much greater than that of Europe. At the present, both remain at a disadvantage relative to the United States. The United States has not grown weaker, rather others including China grow stronger. That growth feeds US paranoia about China which further fuels US resistance to its suggestions. The 2006 DOD report focuses heavily on China's missile forces for example. These views are more intensely held by the George W. Bush administration but not unique to it, at least since the Korean War period.

American military power with its global reach is increasingly dependent upon space assets in order to insure its effectiveness. For example, air strikes employing precision-guided munitions use the US GPS navigation system to direct such weapons to their target regardless of weather conditions. The realization is growing that to stymie the United States, a foe simply needs to take down the US military spacecraft or at least critical components. China's military has openly discussed this apparently weak point in American military power – forces otherwise impossible to counter for most states – what is called asymmetrical warfare. In addition, US commercial comsats and remote sensing satellites are also vulnerable to attack even more easily than the military satellites. For the United States, the question becomes how to protect their satellites, something not thought through well before the 1990s.

Anti-satellite weapons (ASATs) have been possible since the beginnings of the space age. Both superpowers built and tested such systems but in the end found it to their advantage to decommission those systems. For China and other major states, ASATs are attractive because of their capacity to disrupt American military incursions into areas of interest to those states, plus the reality that such weapons can be developed more cheaply than the satellites themselves cost to build or replace. Attacks can also be electronic including jamming, spoofing, laser attacks against receivers, or high energy particle beams. Combined with the slowness with which replacement satellites are sent to orbit, in principle, US threats can be derailed or reduced to impotence. The Pentagon report on Chinese military power cited earlier discusses the question of Chinese ASATs, seeing them as likely to be under development.[15] This perspective had earlier led to a large discussion in reports about Chinese "parasitic" mini-satellites capable of destroying American commercial satellites especially but also military satellites. That was found to be erroneous but it was much discussed for a time, reflecting DOD concern about China's rise to greater stature in the international system and power within the East Asia region.[16]

The US military's response to this scenario has been to harden its military satellites or otherwise render them more difficult to attack through maneuverings in orbit or stealth coating. Maneuvering however shortens the effective satellite life span by exhausting its onboard fuel supply. As a consequence, the DOD has been pursuing refueling satellites in orbit using robotic spacecraft, a logical extension of aerial refueling.

In addition, the US military has proposed placing weapons in orbit for both offensive and defensive purposes, initially as extensions of its ballistic missile defense efforts. American political authorities have not accepted these weaponization proposals. Defensively, these weapons could be used to destroy approaching hostile spacecraft (if identifiable as such) or warheads. One important defensive application first described in the 1980s Strategic Defense Initiative debate – "Star Wars" – used space-based weapons to attack incoming missile buses and warheads. One late concept was the "Brilliant Peebles" – 2,000 small orbiting spacecraft independently directed against ICBMs during their mid-course passage through space. Offensively, such weapons could be used to attack other spacecraft in orbit and possibly strike targets on the ground. The George W. Bush administration in late 2001 formally abrogated the 1972 Anti Ballistic Missile (ABM) treaty.

The ABM treaty had prohibited development of space-based weapons. With its abrogation, the path is now clear to consider all possible ballistic missile defense technologies. For China, ballistic missile defense is truly threatening and destabilizing given its relatively small ICBM forces (see Appendix B). Space-based weapons represent one such technology although once in orbit there is nothing to prevent their use against targets in space other than ICBM warheads and targets on the ground. Thus far, the US government has not moved to weaponization even though the American military assumes that it is simply a matter of time.[17]

For China, that possibility places even more pressure on its space program because all the space applications being pursued, such as remote sensing or reconnaissance, communications and navigation, have the potential to significantly upgrade its military's effectiveness. Ironically, China has now ascended to the level of major space participant just as the international scene reverts back to the early stages of the space age. That is, fundamental questions considered long settled such as the sanctuary concept are now back on the international space agenda. The difference is that the ultimate decision as to how to handle that question will no longer be made by the white man's club of the 1960s and 1970s. Instead, newer space participants, principally China, Japan, India and possibly Europe, will also be central to whatever framework is established. American military power generates a great deal of pressure on other states who feel they are competing politically and possibly militarily with the United States. The George W. Bush administration has fed these concerns with its preemptive military policies.

The American military, especially in its military space circles, considers weaponization as inevitable, which in fact may not be true. For budgetary reasons specifically, and other reasons, the American political leadership may find space weaponization too destabilizing. Technology development will proceed but deployment will likely lag for fiscal and political reasons. Space weaponization is a continuation of the long-standing American quest to substitute technology for manpower.[18] The difficulty is that space technology is prohibitively expensive so that costs at least temporarily derail expanding to include weaponization. This partially explains why the US defense budget, after an initial post Cold War decline, returned to levels commensurate with earlier trends. Fewer personnel are required but much more expensive technologies are employed. This trend predated September 11, 2001.

Starting a space arms race will be both expensive and counterproductive in terms of insuring national security. Placing weapons in orbit will only heighten insecurities because such attacks are virtually unstoppable given the shortened distances and velocities involved. Missile defense involves longer distances and more time than having a space weapon immediately overhead. China's impact is to lengthen the potential competitor list; the comparatively simple bipolar structure of the Cold War is long gone.[19]

Conversely, space weapons are extraordinarily vulnerable in orbit, which returns one to the early nuclear arms race paradox that if you do not use your weapons first, you will lose them (the recipe for preemptive war). This logic led an earlier generation to reject such weapons as destabilizing and likely to bring on the war they were supposed to deter. The one difference that exists presently is that there has been serious consideration of conventional weapons being stationed in orbit, not nuclear, the "rods from god." This reflects the realization that the kinetic energies released from a hypersonic impact may be as effective as a small nuclear blast. Regardless, China's impact on the world security environment can already be seen as its technological capabilities grow.

China's future in space

On occasion, states provide statements indicating the drivers and objectives being built into their space programs. The United States over the years has published a number of space policy pronouncements both general and specific. These general statements usually parcel out responsibilities or delineate future directions for specific areas. Specific policy statements deal with particular applications, such as space transportation or remote sensing. During the Bill Clinton administration in the 1990s, space transportation policy became the subject of one such document. The question driving this particular statement of policy was the conflicting bureaucratic interests of NASA and the US Department of Defense, specifically the Air Force. The policy statement split the task, charging NASA with development of reusable launch vehicles (RLVs) while the military focused on ELVs. Cooperation was possible but not directly mandated.[20] More recently, the George W. Bush administration announced its policy on the question of remote sensing. The issues considered were government control over the field and government support for the private sector.[21]

Chinese space policy from outside China has been deliberately more opaque, in part because the state at least formally remains unified under party control of the government. Rivalries exist within the Chinese space program but the pressures for commercialization are breaking down some old practices. Comparatively little of that bureaucratic struggle flares up in terms of public controversy equivalent to US domestic political struggles. Chinese space activities are the subject of various speeches by party leaders but many earlier statements, especially through the late 1970s, were only partially reported even in China, or their importance was interpreted from vastly different perspectives depending upon evolving political circumstances. As described briefly in Chapters 3 and 4, major political disputes arose in the early years over how quickly the program should proceed. That conflict reflected political concerns about the economic and technological deficits confronting China and disputes arose over how to proceed. More recently, or at least since the early 1980s, the Chinese space program has become an established and valued facet within the government's policy remit. With that general political agreement now firmly in place, it has become easier to formally and publicly articulate a distinctly Chinese space policy.

On November 22, 2000, China issued a White Paper describing the trajectory of China's past, present and future space activities.[22] The importance of this statement is that clearly the Chinese leadership desires to increase public awareness of its space successes, including internationally, unlike earlier years when their fear of failure, national security concerns and domestic political rivalries made such public statements rare. In that sense, the Chinese program mirrored earlier Soviet behavior in that their triumphs were the only items normally reported. Failures advertise to potential enemies evidence of China's weaknesses, increasing the danger level.

Ironically, reporting only successes more likely increased the threat perception as other states feared China's imprecisely understood but growing power. Having limited and partial information fed the already existing paranoia of their military establishments. For example, the United States in the mid-1960s contemplated preemptive strikes against China's strategic assets.[23] In the end, the US decision not to attack was driven by judgments as to the probability of success, considered low due to major intelligence gaps. Likewise, the degree of Soviet secrecy was so successful that there existed great uncertainty about whether the Soviets were actually racing the American Apollo program to the moon. Unequivocal evidence to that point (they were) did not become publicly available until after the fall of the Soviet Union in 1991.[24]

China's growing engagement with the international community, both private, especially commercial, and also governmental, has led to the realization that more transparency is required. The dramatic loss of launch business after the Chinese launch failures in the mid-1990s was exaggerated in part by Chinese demands for secrecy during the ensuing accident investigations – the resulting acrimony aroused suspicions that led to the effective boycott of Chinese launchers by the space insurance industry, completely stripping China's ability to compete in the marketplace. China has taken years to begin recovering its market position, subject to the next launch accident. China's path to recovery competitively has to be built through repeated successful launches of its national payloads. By the spring of 2006, the number of successful launches since 1996 was forty-six – enough for a return to the commercial launch marketplace to no longer be out of the question.[25]

China clearly benefits from the fact that capitalists in principle seek the best launch at the best price, i.e. the cheapest commensurate with conducting a successful launch. This recent focus on price is especially notable since it required the dismantling of the former Cold War security regime that effectively blocked such competition. For years, the launch business remained largely cost-insensitive due to security concerns; now costs matter even for national militaries, although they are still likely to pay a premium if necessary.

At the present time and for the immediate future, space commerce, especially the launch sector, is a highly competitive business, especially given the continuing paucity of large payloads against an expanding number of launcher options. Smaller payloads (less than 100–300 kg) are generally launched by national flag carriers if they exist because the cost to fly internationally for a single payload is too expensive given the value of the payloads. Exceptions occur, however, when multiple payloads fly on the same booster. For example, the Iridium comsat flotilla consisting of sixty-six comsats carried to orbit in varying size clusters on the same rocket. Long March rockets on September 1 and December 8, 1997 carried two comsats at a time to orbit, while the Russian Proton rocket earlier carried up to seven at a time.[26]

The White Paper being discussed here was released prior to the successful launch of the Shenzhou 5 to orbit and back in October 2003, which means the document reflects official Chinese views articulated prior to the great surge in public enthusiasm that occurred in China after that first manned mission. The document itself is short, slightly over twelve pages in English, but provides some interesting insights into the underlying principles and future directions of the Chinese space effort. Their general approach can be summarized in terms of Chinese aims along with the principles underlying those aims. The specific aims enunciated in this White Paper reflect the peculiarities of Chinese space goals and experience, including its strong emphasis on national economic and technological development.

> The aims of China's space activities are: to explore outer space, and learn more about the cosmos and the Earth; to utilize outer space for peaceful purposes, promote mankind's civilization and social progress; and benefit the whole of mankind, and to meet the growing demands of economic construction, national security, science and technology and social progress, protect China's national interests and build up the comprehensive national strategy.[27]

China's statement with its evocation of the exploration aim sounds very traditional – the exploration theme has been in all such general statements from a variety of states. The key forces that actually drove Chinese space activities come at the end with the emphasis on "economic construction, national security ... [and] protect[ing] China's national interests." Space activities are not considered a luxury for the Chinese leadership, rather the question is what those critical activities can bring to enhance China's economy and defense capabilities. As has been indicated previously, each decision to move to the next level of space activity encountered opposition since those precious resources were needed elsewhere. From those aims, the following five principles are deduced:

1. Adhering to the principle of long-term, stable and sustainable development and making the development of space activities cater to and serve the state's comprehensive development strategy.
2. Upholding the principle of independence, self-reliance and self-renovation and actively promoting international exchanges and cooperation. China shall rely on its own strength to tackle key problems and make breakthroughs in space technology.
3. Selecting a limited number of targets and making breakthroughs in key areas according to the national situation and strength. China carries out its space activities for the purpose of satisfying the fundamental demands of its modernization drive.
4. Enhancing the social and economic returns of space activities and paying attention to the motivation of technological progress. China strives to explore a more economical and efficient development road for its space activities so as to achieve the integration of technological advance and economic rationality.
5. Sticking to integrated planning, combination of long-term development and short-term development, combination of spacecraft and ground

equipment, and coordinated development. The Chinese government develops space technology, application and science through integrated planning and rational arrangement in the aim of promoting the comprehensive and coordinated development of China's space activities.

Within the White Paper itself, there are repeated references to China's space successes but their recognition is hedged with caveats and it is emphasized that the space program excels in selected but "important fields of space technology." The major areas of successful Chinese technological development in their judgment include: "satellite recovery, multi-satellite launch with a single rocket, rockets with cryogenic fuel, strap-on rockets, launch of geostationary satellites and TT&C (Telemetry Tracking and Command)."

The above list of aims and principles boils down to two major themes: achieving long-term stable and sustainable economic development while employing space applications as one critical tool; and maintaining Chinese political and economic independence from outsiders while continuing to grow in national capabilities and strength. In those principles, especially number 3, there is clear recognition of China's continued need to focus its efforts and resources on selected areas. Resources are not available for dramatic increases in funding across the board. Cooperative activities with other states and multinational corporations are one means by which China can continue and even accelerate its technology development across a wider spectrum of such space activities.

The difference is that China now feels it approaches cooperative activities from a general position of strength, an equal in many respects to the strongest states and superior to many others, no longer an inferior. One can see that confidence in China's repeated suggestions that it join the ISS program in some fashion. From China's perspective, the Shenzhou program is building the requisite technologies that facilitate such cooperation. In April 2006, Luo Ge of the China National Space Administration visited NASA and invited NASA Administrator Michael Griffin to visit China.[28] An earlier congressional delegation to China had suggested development of a compatible docking system so that the Shenzhou could link up with the US Crew Exploration Vehicle (CEV). All of this follows Chinese proposals made the year before that the Shenzhou be used as a cargo/taxi vehicle for the ISS, supplementing the Russian Soyuz and Progress vehicles, the US space shuttle until finally shut down around 2010, the European Automated Transfer Vehicle (ATV) still under development, and the US CEV (under development). In fact, the two Russian vehicles and the Shenzhou were the only flight systems considered operational at the time of the April 2006 meeting.[29]

What is important is not the details, which are still in flux, but the fact that China has grown sufficiently confident to extend the offer. For the government, this represents further validation of their own view of Chinese capabilities. So, in that sense, China has clearly emerged from the ranks of the underdeveloped world to world heights in this particular technological arena. This perception, however, is tempered by the economic challenges that remain.

International cooperation can assume many forms, the ISS being only one, albeit the most visible and prestigious despite the program's persistent problems. For China, the United States presently holds the keys to the door regarding Chinese participation, and the United States remains focused on the role of the PLA in controlling the space program and proliferation issues generally.[30] Chinese overtures to the US continually confront that American concern that active Chinese participation would enhance its military sector.[31]

China therefore has moved toward greater cooperation with other partners, especially the Europeans, as evidenced by China's participation in the Galileo navigation satellite program.[32] The Europeans unlike either the Russians or the Americans have been engaged in truly cooperative space activities almost from the inception of their space efforts. The Russians and Americans started as solo practitioners of space activities – a status still clung to most of the time.

Not until the ISS program stalled in the early 1990s did the United States finally begin participating cooperatively with others. That shift was symbolized by the presence of both the Russians and Canadians in the critical pathway to ISS completion.[33] (Critical pathway refers to those steps that have to be completed if the program is to succeed.) Previously, the relationship had normally been one of superior to subordinate. American cooperation with China in civil space activities is inevitable, barring some total breakdown in their relationship, most likely over Taiwan.

Four pages of the White Paper discuss what China has accomplished over the past forty years of its space activities, especially the past two and half decades. Those efforts have been described earlier and require no repeating here. What is more interesting is their short- and long-term development targets. In the short term (over the decade from 2000), the goals are partly prosaic and eminently achievable. Building on their existing work, remote sensing, navigation and telecommunications along with improved launch capabilities are the major goals being pursued in terms of concrete programs.

Human spaceflight and space science are also included but the heaviest emphasis remains focused upon the economic developmental potential of space applications. By contrast, the two pioneer space participants pursued the military and political first, with economic developmental aspects temporarily obscured by those national security drivers, although developmental activities occurred, especially for the Americans. For the United States, the commercial-developmental aspects moved to the fore reasonably quickly, while for the Soviets, the military-political aspects remained dominant much longer, effectively until the collapse of the Soviet Union.[34]

Instead, China's long-term goals include greater commercialization of their expanding space applications along with an expanded human spaceflight program, with space science carried out on an even more intense level. What is interesting is the fact that commercialization is not an explicitly stated primary concern in the short-term goals. That absence reflects the Chinese government's awareness of the issues raised about the technical quality of Chinese satellites. Until recently, Chinese satellites were too short-lived once they reached orbit to be truly competitive in the commercial space marketplace. Satellite replacement represents a major expense for commercial enterprises. The Soviets had suffered the same problem, one factor explaining why they launched twice as many replacement satellites as the United States. This technology quality problem is being overcome but first impressions of earlier Chinese satellites linger and hinder their immediate market prospects. Effective in-orbit life spans of under five years cannot compete commercially with projected twelve- to twenty-year life spans of major US and European competitors. China's satellite industry currently survives on the basis of domestic satellite orders from the government – Chinese-based telecommunications companies purchase European comsats – and, formerly, purchased US comsats before the security restrictions grew too onerous. Improving quality remains difficult, as the Japanese have discovered with repeated failures of their spacecraft.[35]

As discussed in Chapter 5, on October 15, 2003, the first Chinese yuhangyuan or taikonaut, Lt. Col. Yang Liwei, flew to orbit and returned safely. For China, this event was galvanizing – certifying in the Chinese public eye the ascension of China to the highest international levels of space activities. This event fed Chinese nationalism – in the rush of enthusiasm, public expectations concerning Chinese space plans soared. Reality proved more modest with China flying its second manned mission in October 2005, two years after the first flight. The delay reflected Chinese caution since their leaders abhorred the prospect of public failure, especially since global attention was now clearly focused on their efforts. Transparency becomes a two-edged sword as the American space program discovered with its several shuttle flight failures. There is nowhere to hide although one basks in the glory.

Regardless of their caution, China even before the Shenzhou 5 flight announced a series of proposed missions both manned and robotic. With their new political and program momentum, the human spaceflight side of the program draws much speculation, especially since many outside China assume they will push forward as rapidly as the Soviets and Americans did in the 1960s. Such a view ignores the greatly changed circumstances under which China's program presently operates. The Chinese presently race alone – the two space pioneers have been there and done that, while any others lag far behind.

There are those, especially in the United States, who speculate that the United States and China are engaged in a new space race.[36] This concern by some congressional members led one congressional committee to demand an unclassified report from NASA on China's space program.[37] The fact is that up to now proponents of a space race have not generated any appreciable political traction. In fact, some in the US Congress argue that China should be brought into the ISS program.[38] Other potential competitors, India and Japan primarily, lag far behind China and are not likely to dramatically accelerate their human spaceflight programs for another decade. The Europeans are currently in similar circumstances.

Pitfalls on the road to the future

China's space program, Program 921, prior to the events of October 15, 2003, envisioned a steady progression – in effect, a building block or modular approach. That approach has served China well because it has kept expectations more or less under control. After Shenzhou 7, however, the political pressures to accelerate the program or not may grow more intense. At that point, the national leadership must calculate anew how much is enough to sustain the program and continue advancing.

Human spaceflight is the great adventure but given China's economic and social goals, hard economic choices must be made. Both the United States and Russia eventually had to calculate those same values and in effect scale back their human spaceflight programs. The Americans, especially with their Vision for Space Exploration, confront the harsh realities of fiscal and resource constraints most publicly, but Russia is also constraining its program.

China has laid the groundwork for success but now the truly heavy lifting begins as the technologies and aspirations both grow even more expensive. Staying the course will require strict economic and political discipline by both the space program leadership and the government. The temptation will be to push the space program, especially the Shenzhou component, forward even more quickly. The nationalist pride of the Chinese people was obviously aroused after the October 15, 2003 Shenzhou 5 flight. Rekindling those intense emotions may be important for a regime confronting potentially severe internal dislocations due to its continuing economic transition. Although some complain that the space program has squeezed resources that could have been used somewhere else, the overall leadership support increased with the success of the Shenzhou 5 flight. Given the fact that bread is not an issue anymore and the private sector sustains China's economic development, political elites do not need to think too much about bread and circuses but rather follow any government's normal reflex to try to garner increased public support through invoking nationalist sentiments.

For both the space pioneers, similar nationalist feelings of great pride were generated but after the initial euphoria when their programs became more normalized those feelings declined (but did not disappear) in intensity.

At that point, the state space programs declined as a national priority, meaning slowed or flat budgetary growth – a normal response historically speaking. For NASA, 1 percent of the federal budget is the figure usually cited as evidence of serious support or not.[39] This view usually references the anomaly of the Apollo program in the 1960s as the gold standard, when the percentage reached 4.4, never matched since that time. For China, that transition will likely come after 2010 when the Shenzhou program can be considered firmly established. The Moon lies beyond but the true price may be excessive for the benefits received; then the cooperative option may become much more attractive. The Americans already confront that question; the benefits are largely intangible for the state even though certain contractors will benefit individually from building the equipment and supporting the program. National leaderships must confront this equation each time such expensive undertakings appear on the national agenda. Accurate budget figures on the Chinese space program are scarce; up to this point the costs have been managed successfully, but the political pressures grow more intense to remove the budget limits.

Four major issues confront China from this point forward: first, how stable the Chinese situation is in terms of continuity regarding space policy; second, what stance to take regarding international cooperation; third, transitioning from a government-dominated to a mixed space program; and fourth, future directions regarding military space activities. All of these questions will impact the future directions and successes of the Chinese space program.

First, although the present strong government commitment to space activities and fast development seem to imply a bright future for the Chinese space program, China's ultimate fulfillment of these plans is not guaranteed. There are several reasons. First, irrational policy making has repeatedly hampered the sustained development of the Chinese space program. This study shows that the motivations supporting the Chinese space program have varied significantly with changes in the domestic political interests of the paramount leader. The combination of centralized power and an inactive role by ordinary citizens leads to political instability, with one consequence being wide policy swings. This problem also plagues Chinese space policy. More than once, government space plans have been cancelled or dramatically altered. Since the retirement of Jiang Zemin, China appears to have entered a more stable and institutionalized power transition process, but no one can guarantee that the irrationality found in Mao's era will never happen again. Thus, whether the ambitious space plans made since 2004 can in fact be carried out remains a problem, since no one can guarantee that they will not be cancelled by a future leadership. The economic boom and the present fast overall development of science and technology have laid a necessary foundation for China to carry out its ambitious programs, but political uncertainty might put the Chinese space program in reverse or stall further progress. There is no definitive answer to this question.

Second, the question of international cooperation comes in two forms: China's current practice and entering into an international leadership role. Currently, China is engaged in cooperative space programs with a number of states including a proto-Asia-Pacific equivalent to ESA. The Asia-Pacific Space Cooperation Organization (APSCO) established in 2005 is an excellent example of China working with states for whom China is the leader. The original APSCO members include: Bangladesh, Indonesia, Iran, Mongolia, Pakistan, Peru and Thailand, along with China.[40] Its bilateral agreements are often with states in relation to whom China is clearly the dominant partner. Whether APSCO evolves into something more is unknown but China is clearly the giant among space pygmies.[41] As a consequence, China is in charge of how fast and in what direction the particular partnership moves over time. In this situation, the Chinese are in effect the Americans or the Russians in their heyday.

The other side of the cooperation coin is whether China should aggressively pursue its role as potential international partner-leader for large-scale projects. China currently pursues such a course, as occurred for example during President Hu Jintao's April 2006 visit to the United States. An earlier invitation to NASA administrator Michael Griffin had been deferred but became a higher priority when President Bush indicated that a visit by Griffin in the fall of 2006 was now appropriate.[42] In addition, the United States and China signed protocols extending nine bilateral S&T cooperative agreements.[43] At the same time, the China National Space Administration was invited (at the last minute) to participate in an Exploration Strategy Workshop. This workshop was in support of the US Vision for Space Exploration in which the United States is attempting to set the agenda for the next big wave in human space exploration. Its potential partners appear more skeptical of US leadership and are pushing for major changes before they finally commit to the program. Chinese participation in the workshop for some reason was brief – the Chinese left before the break-out sessions.[44]

China must be careful what it wishes for with regard to international space cooperation. International cooperation agreements are normally sold to the political leadership as having great value politically and otherwise. The reality is that large-scale space projects can turn into financial albatrosses which grow relentlessly. Terminating the project when costs explode would seem an easy decision but, in reality, notions of national prestige and pride prevent such a decision from being made. Such programs are best strangled at the first sign of trouble because, once underway, domestic constituencies develop who resist change in order to protect jobs created by the program. In a developing society, jobs are critical, especially high-tech ones which provide employment for younger workers. Instead, more funds are committed because, it is argued, other states depend on your continued participation. Leading such projects may fill a state's leadership with great pride but the downside comes when the bills come due and the state finds itself unable to extricate itself from the program. The classic example is the ISS program where the United States and the original partners, Europe, Japan and Canada, felt obligated to remain even as the program morphed into a budget eater. For reasons of international leadership, the US was never willing to admit failure and cut its losses.

Entering into the ISS program and the Galileo navigations satellite program is easy for China presently because the heavy lifting financially is being done by the original partners. The ISS in different guises has been underway officially since 1984, while the Europeans want to control Galileo so international partners are welcome but their roles are consciously limited. When Galileo's military applications become more public, the European Union will further limit the influence of outsiders. The space programs of Europe, Japan and Canada have all had their space program budgets and agendas at times dominated by their commitments to completing the ISS. Getting the ISS built has turned into a death march across multiple administrations and governments to arrive at a product that in the end does very little of what was originally proposed.[45] The United States engaged in repeated redesigns of the ISS, usually with minimal or no prior discussion with the partners. Now, close to success in terms of completion (albeit a much scaled-down version), the ISS program confronts its demise almost before it is finished. There is much criticism of US leadership in this program but the long-standing American retort is that no other state was willing to step up and lead while paying most of the bills. Ironically, NASA used the international partners as the justification for continued US participation and financial support; canceling was seen as extremely embarrassing.

Becoming an important international leader among advanced space states carries with it the implied burden of meeting the bills and providing technological leadership across the board. To this point, China's program has reached success by picking its spots in terms of programs and what budget allocations are made. For years, the US government was continually pressured to do more financially than it wished to do, under the guise that to not do more signaled a failure of US leadership. China's present status with regard to most of its current partners is clearly that of leader – a position easily maintained when the relevant partners are significantly weaker. Adding Japan to APSCO would change the dynamics since Japan if so motivated can compete with China with regard to space technologies and their applications. For China, the decision as to how to work with others will be crucial for keeping its space program on track and within the budget the leadership desires. This is not a trivial point since programs can become overcommitted, which detracts from their central focus. China for the past three decades has prospered with regard to its space efforts because of that focus and discipline, which is easier when others' impact is minimized. Becoming more engaged limits that flexibility and unity of purpose.

In the Chinese case, another important question becomes how fast the transition will be from its current completely government-dominated space program to a mixed one similar to those of the two space pioneers and other advanced states. What triggers this transition for China is the question of how large a role the market will have in operating Chinese space technologies. The space program in China to a large extent reflects the old planned socialist economy, in which the government makes the plans, state-owned R&D institutions and companies carry out the plans, and then the products go back into the hands of government. The planned economy is notorious for its low efficiency levels and huge waste of resources. China's overall economic scale in the twenty-first century is much larger than that of the Americans in 1959. But in 1959, the United States launched thirty-seven satellites. Has China done that? The answer is no. The total number of satellites that China planned to launch in the five years from 2000 to 2005 was only thirty-five. An insightful article in *Nature* points out that heavy government involvement may lead to R&D failures since both administrators and researchers are likely to use such research to advance their own political or economic goals.

> Close government involvement without an adequate external review system, could lead to fruitless projects and could even stall scientific development. The risk is that big projects will become an end in themselves, where both politicians and scientists demonstrate their power by the size of budget they acquire for their project, regardless of its scientific merit.[46]

Seed experiments might be one example from the space field. More than once, China launched seeds into Earth orbit; subsequently, Chinese scientists claimed that the seeds exposed to cosmic radiation produced higher yields. But the suspicious side of their success story is that NASA has launched seeds into orbit but no dramatic successes have been reported, despite the fact that the Americans have been actively searching for ways to utilize outer space to increase productivity including the agricultural output. If there was a possibility, it would have been pursued. The explanation here possibly lies in the "power demonstration" factor pointed out by *Nature*.

The Great Wall Industry Corporation presents the image of privatization to the world but the reality is total government control rather than market judgments. The Corporation's function is to facilitate government policy; profits are an additional incentive but subordinate to the political. But, as China competes in the growing international marketplace, the question will rise to the fore: why should the government fund and operate technologies that can be operated for a profit? In the American context, the argument was against corporate welfare, while in Russia's case, the goal became economic competitiveness since the Russian government in the 1990s lacked the resources to fund its space industry in the style it had become accustomed to during the Soviet era. China is reaching the point where such a question will become politically potent given the present economic proclivities of the government in other sectors. The key will be whether the Chinese private space sector is able to free itself from government control and grow economically. Running a state organization for real profit will be severely tested if Chinese space technologies are truly going to be competitive worldwide. Cost and exchange rate differentials will begin to narrow, demanding even greater efficiencies. The failure to adjust will be socially and economically disruptive, not a desired condition for the government.

China's well documented problems with its state-owned enterprises illustrate the need to change the expectations and culture within such organizations.[47] Change is coming and it will eventually expand to incorporate the space sector. Change, however, will not be immediate, in large part because the entire field remains immersed in government perspectives, especially regarding the launch area with its dual-use implications. Other areas of space applications are becoming commercialized, however. The old government telecommunications regime internationally has ended, witness the demise of Intelsat and Inmarsat, the quasi-government consortiums that controlled international satellite-based telecommunications from the 1960s onward. That change process has also reached the domestic space industry in many states; China will eventually reach the same stage unless there is a major shift in national economic policy.

Finally, military space activities represent the wild card for China, as they do for every state that aspires to international leadership. In the beginning, the military implications were an overriding priority because demonstrating space access also meant a viable missile force existed or was possible. Once that image or reality was established, China's interest in military space waned, as did the space program generally until the 1980s. What happened was that China was left far behind militarily as the Soviets and the Americans in particular developed the full spectrum of military satellite applications in order to enhance their military effectiveness and efficiency. Those applications have recently been augmented by the employment of commercial satellites for remote sensing and communications, expanding the US military's coverage of the world. For example, US military operations in Afghanistan in 2001 relied heavily on commercial satellites.

For China and other states, the US military operations in Kosovo, Afghanistan and Iraq demonstrated the fact that a military employing the full repertory of military space applications can dominate the battle space. American forces demonstrated great agility and flexibility in conducting combat operations in remote sites. For states such as China, enhancing those capabilities becomes a necessity given China's concerns about Taiwan and Japan. American operations in Iraq in 2003 demonstrated both the effectiveness and the limits of this new warfare. The United States quickly defeated the degraded Iraqi military but its long-term capacity to deal with other aspects such as occupation and counter-insurgency operations is in doubt.

The importance for China's space program comes in the heightened demands placed upon its budget and how that is distributed across the spectrum of space activities. With states such as India (a regional rival) openly espousing the concept of military space operations, the pressures will mount in support of hiking the military side of the equation.[48] This could dramatically slow the Chinese human spaceflight program and the scientific missions aimed at the Moon and possibly other celestial bodies. The focus will narrow to those aspects of the space program most supportive of national security. The crunch will come because China despite its rapid economic growth is still a developing country with the needs and limitations that implies. China's government has become supportive of space activities for a variety of reasons but principally in support of the state's survival. Originally, that was interpreted in terms of developing the missile forces considered necessary for protecting China. Next, the focus shifted to economic development but the kernel that lay at the heart of both efforts was supporting China as it made its way in an often hostile world. That will continue to be the motivation driving China's progress up the stairway to heaven.

Conclusion

China's space program was a journey that began in the 1950s when it took the first tiny steps to acquire ballistic missiles which had the potential to also allow China to enter the heavens above. The journey was neither quick nor easy but one in which perseverance paid off in the end. The difficulties were both natural and human; the first proved easier to conquer than the second but progress was made and continues to the present. China stands astride the divide between the third and first worlds of economic development. That inspires some and frightens others but the reality is that China is now a player in the international space policy arena.

Appendix A

Chinese launch vehicles

Launch vehicle	LEO	SSO	GTO	First flight	Builder
CZ-1	300	–	250	1970	CALT
CZ-1C	500	–		1988	CALT
CZ-1D*	740	300	440	1995	CALT
FB-1	2,500	–	–	1972	SAST
CZ-2	1,800	700	–	1974	CALT
CZ-2A	2,000	–	–	1974	CALT
CZ-2C*	2,500	700		1975	CALT
CZ-2C/SD	2,800	700	–	–	CALT
CZ-2D*	3,500	–	1,205	1992	CALT
CZ-2E*	9,200		3,370	1990	CALT
CZ-2E(A)*	14,100		3,375	2000	CALT
CZ-2ELA	12,000		–	–	CALT
CZ-2F/manned*	8,400		3,500	1999	CALT
CZ-3*	4,800		1,400	1984	CALT
CZ-3A*	7,200	–	2,700	1994	CALT
CZ-3B*	11,200		5,100	1995	CALT
CZ-3B(A)	13,000		6,000	2002	CALT
CZ-3C		–	3,700		CALT
CZ-4A	4,680	1,650	1,100	1988	SAST
CZ-4B		–	1,500		SAST
CZ-5	25,000				CALT
KT-1	100			2002	Space Solid Fuel Rocket Carrier
#CZ-NGLV-200	1,500			2008	CALT
#CZ-NGLV-320	10,000		6,000	2009	CALT
#CZ-NGLV-522/HO			11,000	2010	CALT
#CZ-NGLV-504/HO			14,000	2011	CALT
#CZ-NGLV-540.HO			6,000	2012	CALT
#CZ-NGLV-504	25,000			2013	CALT
#CZ-NGLV-522	20,000			2015	CALT
#CZ-NGLV-540	10,000			2016	CALT

CZ – *Chang Zheng* (Long March)
FB – *Feng Bao* (Storm)
KT – *Kaituozhe* (Pioneer)
SAST – Shanghai Academy of Spaceflight Technology
CALT – China Academy of Launch Vehicle Technology

* in service
\# future development

The list above was constructed from various references; the Chinese like other states have built multiple variations of their launch vehicles often to cover very specific needs. At the bottom of the table, the projected new launch vehicles are listed with estimated times of development. The list is logical and sequential but will obviously change over time as priorities change and developmental questions arise. As one can see, China now has big dreams regarding its space programs; the new launch options are in pursuit of those dreams.

Appendix B

Chinese strategic missiles

Missile designation	First launch	Range (km)	Type	Fuel type
DF-1	1960	590	IRBM	Liquid
DF-2 (CSS-1)	1962	1,050	IRBM	Liquid
DF-2A (CSS-1)	1970	1,250	IRBM	Liquid
DF-3 (CSS-2)	1966	2,810	IRBM	Liquid
DF-4 (CSS-3)	1970	4,760	IRBM	Liquid
DF-5 (CSS-4)	1971	12,020	ICBM	Liquid
DF-57	1978	6,900	ICBM	Liquid
DF-21 (CSS-5)	1988	1,810	IRBM	Solid
DF-15 (CSS-6)	1990	605	IRBM	Solid
DF-31 (CSS-X-9)	1998	7,900	ICBM	Solid
DF-41 (CSS-X-10)	2010	12,900	ICBM	Solid

DF – *Dong Feng* (Chinese terminology; the United States uses the CSS designator for Chinese missile systems)

China's missile program continued in parallel with its civil space program, as can be seen in the above table. The major steps were lengthening the effective ranges of the missiles and shifting to solid fuel from liquid. The latter makes the missiles more immediately responsive, plus the fuels are more stable over long time spans. Nuclear warheads were developed but their deployment was slow for many years as the Chinese leadership grew more confident that the United States was not likely to preemptively attack China and the Soviet Union collapsed. This reduced the pressures to quickly implement a large ICBM force; rather the focus was on intermediate range ballistic missiles (IRBMs), especially deployed opposite Taiwan as a deterrent to its aspirations for political independence.

Notes

I Overcoming the past, seizing the future

1 William D. Wray, "Japanese Space Enterprise: The Problem of Autonomous Development," *Pacific Affairs* 64 (Winter 1991–1992), 463–488. Japan's effort at establishing an autonomous space capability ran into intense resistance from the Americans who perceived Japan's efforts as another phase in their effort to dominate specific industries as they already did in electronics.
2 Marcia S. Smith, *Potential International Cooperation in NASA's New Exploration Initiative* (Washington: Congressional Research Service, April 27, 2004), CRS-1.
3 *Ibid.*
4 "Brazil's Difficult Road to Space," *Space Today Online* (August 23, 2003), http://www.spacetoday.org/Rockets/Brazil/BrazilRockets.html (accessed May 29, 2006).
5 Interview, "John Lewis Gaddis on: The Role of the Atomic Bomb in the Korean War," http://www.pbs.org/wgbh/amex/bomb/filmmore/reference/interview/gaddis5.html (accessed May 17, 2006); Philip Short, *Mao: A Life* (New York: Henry Holt, 1999), 489–490.
6 Roger Handberg, *Seeking New World Vistas: The Militarization of Space* (Westport, CN: Praeger, 2000), 33–57.
7 Frank Morring, Jr., "Program of Programs," *Aviation Week & Space Technology* (November 22, 2004), 34.
8 Roger Handberg, "The Future of American Human Spaceflight: Bureaucratic Visions and Political Realities, the Vision for Space Exploration." Paper presented at the Annual Meeting of the Southern Political Science Association, New Orleans, January 2005.
9 For one analysis, see Joan Johnson-Freese, "'Houston, We Have a Problem': China and the Race to Space," *Current History* (September 2003), 259–265.
10 William E. Burrows, *This New Ocean: The Story of the First Space Age* (New York: Modern Library, 1998), 421–428.
11 Chairman of the Joint Chiefs of Staff, *National Military Strategy of the United States of America* (Washington: Office of the Joint Chiefs of Staff, 2004).
12 Department of Defense of the United States, *Quadrennial Defense Review Report* (Washington: US Department of Defense, September 30, 2001 (accessed October 2, 2001).
13 Financial Times, "US signals hard line on China," *Financial Times* (February 21, 2005). 2006 Quadrennial Defense Review Report (Washington: Office of the Secretary of Defense, February 6, 2006).

14 Roger Handberg, *International Space Commerce: Building from Scratch* (Gainesville, FL: University Press of Florida, 2006).
15 David J. Whalen, *The Origins of Satellite Communications, 1945–1965* (Washington: Smithsonian Institution Press, 2002).
16 Douglas Barrie, "Strategic Space," *Aviation Week & Space Technology* (November 15, 2004), 77.
17 Joan Johnson-Freese, *The Chinese Space Program: A Mystery Within a Maze* (Maibar, FL: Krieger Publishing Company, 1998).
18 Brian Harvey, *The Chinese Space Programme: From Conception to Future Capabilities* (New York: Wiley, 1998).
19 Brian Harvey, *China's Space Program: From Conception to Manned Spaceflight* (New York: Springer-Praxis, 2004).
20 Mark Stokes, *China's Strategic Modernization: Implications for the United States* (Carlisle, PA: Strategic Studies Institute, 1999).
21 Frank Morring, Jr. and Neelam Mathews, "Third World Rising," *Aviation Week & Space Technology* (November 22, 2004), 44–49.
22 Frank Morring, Jr. and Neelam Mathews, "Application Driven," *Aviation Week & Space Technology* (November 22, 2004), 48–50.
23 Roger D. Launius, "After *Columbia*: The Space Shuttle Program and the Crisis in Space Access," *Astropolitics* 2 (Autumn 2004), 308.
24 Walter McDougall, . . . *The Heavens and the Earth: A Political History of the Space Age* (Baltimore, MD: Johns Hopkins Press, 1985); Graham T. Allison, *Essence of Decision: Explaining the Cuban Missile Crisis* (Boston: Little, Brown, 1971).
25 "China Gets the Bomb, Statement of the Government of the PRC, October 16, 1964," *Modern History Sourcebook*, http://www.fordham.edu/HALSALL/mod/1964china-bomb.html (accessed May 27, 2006).
26 Robert A. Divine, *The Sputnik Challenge: Eisenhower's Response to the Soviet Satellite* (New York: Oxford University Press, 1993).
27 Harvey, *The Chinese Space Program*, 39.
28 Harvey, *China's Space Program*, 70–73.
29 Thomas J. Christensen, "China, the US–Japan Alliance, and the Security Dilemma in East Asia," *International Security* 23 (Spring 1999), 49–50. For broader discussions, see Robert Jervis, "Cooperation under the Security Dilemma," *World Politics* 30 (January 1978), 167–174; and Jervis, *Perception and Misperception in International Politics* (Princeton, NJ: Princeton University Press, 1976), ch. 3.
30 Roger Handberg, *Ballistic Missile Defense and the Future of American Security* (Westport, CN: Praeger, 2002), 56–58.
31 Craig Couvault, "Iran's 'Sputnik,'" *Aviation Week & Space Technology* (November 29, 2004), 36–37.
32 John Kelly, Chris Krindler, and Kelly Young, "$1 Billion Question: Where Has All the Air Force EELV Money Gone?" *Florida Today* (August 22, 2002), http://www.space.com/missionlaunches/fl_eelv_020825a.html (accessed May 23, 2006).
33 Rip Bulkeley, *The Sputniks Crisis and Early United States Space Policy* (Bloomington, IN: Indiana University Press, 1991).
34 Professor J.L. Lions, *Ariane 5, Flight 501 Failure Report by the Inquiry Board (July 19, 1996)*, http://java.sun.com/people/jag/Ariane5.html (accessed November 5, 2003). Success followed: Peter B. de Selding, "Ariane 5ECA Launch Boosts Arianespace's Competitiveness," *Space News* (February 21, 2005), http://www.space.com/spacenews/archive05/Ariane-0221.html (accessed June 21, 2005).

35 Missile Technology Control Regime, http://www.mtcr.info/english/index.html (accessed January 17, 2006).
36 Short, *Mao*, 425–533.
37 Li Shenzhi, "When Did Mao Zedong Decide to Lure Snakes Out of Their Holes?" *Works of Li Shenzhi*, http://members.lycos.co.uk/sixiang000/author/L/LiShenZhi/LiShenZhi008.txt (accessed June 9, 2006).
38 Lowell Dittmer, *Liu Shao-ch'i and the Chinese Cultural Revolution: The Politics of Mass Criticism* (Berkeley, CA: University of California Press, 1974).
39 Leonid Brezhnev, "Speech before Fifth Congress of the Polish United Workers' Party," Warsaw, Poland, November 13, 1968.

2 China as space follower and leader

1 Stan Lehman, "Space Program Leaders Blamed for Brazil Accident," *USA Today* (March 16, 2004), http://www.usatoday.com/tech/news/2004-03-16-brazil-report_x.htm (accessed January 1, 2005).
2 Stephen B. Johnson, *The Secret of Apollo: Systems Management in American and European Space Programs* (Baltimore, MD: Johns Hopkins University Press, 2002); Asif A. Siddiqi, *Sputnik and the Soviet Space Challenge* (Gainesville, FL: University Press of Florida, 2000).
3 C.N. Hill, *A Vertical Empire: The History of the UK Rocket and Space Program, 1950–1971* (London: Imperial College Press, 2001); Michelangelo De Maria, *The History of the ELDO, Part 1: 1961–1964* (Noordwijk, Netherlands: European Space Agency, HSR-10, September 1993).
4 Roger Handberg and Joan Johnson-Freese, *The Prestige Trap: A Comparative Study of the United States, European, and Japanese Space Programs* (Dubuque, IA: Kendall/Hunt Publishing, 1994), 95–99.
5 Roger Handberg, Joan Johnson-Freese and George Moore, "The Myth of Presidential Attention to Space Policy," *Technology in Society* 17 (1995), 337–348.
6 Roger Handberg, "Dancing with the Elephants: Canadian Space Policy in Constant Transition," *Technology in Society* 25 (2003), 27–42.
7 David Callahan and Fred I. Greenstein, "The Reluctant Racer: Eisenhower and US Space Policy," in Roger D. Launius and Howard E. McCurdy (eds), *Spaceflight and the Myth of Presidential Leadership* (Urbana, IL: University of Illinois Press, 1997), 15–50.
8 David A. Fulghum, "US Doubts Korean Space Launch Claim," *Aviation Week & Space Technology* (September 14, 1998), 58–59; Roger Handberg, "Changing Parameters of Japanese Security Policy: The Advent of Military Space in the Post Cold War Environment." Paper presented at the Annual Meeting of the Southern Political Science Association, Savannah, November 1999.
9 Geoffrey Forden, "The Military Capabilities and Implications of China's Indigenous Satellite-Based Navigation System," *Science and Global Security* 12 (2004), 219–250. The purpose of the Beidou appears clear, enhancing China's strategic forces.
10 Lorenza Sebesta, *The Availability of American Launchers and Europe's Decision "To Go It Alone"* (Noordwijk, Netherlands: European Space Agency, HSR-18, September 1996).
11 "Missile Technology Control Regime," http://www.mtcr.info/english/ (accessed March 17, 2006).
12 Roger Handberg, *International Space Commerce* (Gainesville, FL: University Press of Florida, in press), ch. 4.

13 *Columbia Accident Investigation Board Report: Volumes I–VI* (Washington: NASA and Government Printing Office, August 2003).
14 Tarq Malik, "Orbital Rendezvous: Fresh Crew, Brazilian Astronaut Arrive at ISS," *Space.com* (April 1, 2006), http://www.space.com/missionlaunches/060401_exp13_docking.html (accessed April 3, 2006).
15 For the official description, see European Space Agency, "Crew Return Vehicle," http://www.esa.int/esaHS/ESARZS0VMOC_iss_0.html (accessed February 12, 2006).
16 Marcia Smith, "Space Exploration: Issues concerning 'Vision for Space Exploration,'" *CRS Report to Congress* (January 4, 2006).
17 Jeff Foust, "Small Steps Forward for New Space," *The Space Review* (April 24, 2006), http://www.thespacereview.com/article/609/1 (accessed April 25, 2006).

3 First awakenings

1 Walter A. McDougall, . . . *the Heavens and the Earth: A Political History* (New York: Basic Books, 1985, reprinted Baltimore: Johns Hopkins Press, 1997).
2 Yanping Chen, "China's Space Policy – a Historical Review," *Space Policy* (May 1991), 116–128, http://www.cast.ac.cn/gycast/ls.htm
3 The analysis here obviously draws heavily from the research by Brian Harvey. His work has been reported most fully in two volumes, in effect, his first and second editions. The relevant volumes are: *The Chinese Space Programme: From Conception to Future Capabilities* (Chichester and New York: Wiley, 1998); and *China's Space Program: From Conception to Manned Spaceflight* (New York: Springer-Praxis, 2004). In addition, work done by Joan Johnson-Freese as reported in her book, *The Chinese Space Program: A Mystery within a Maze* (Melbourne, FL: Kreiger Publishing, 1998); and various articles including "Strategic Communications," *China Security* (2006), 37–57; "China Bids for High Ground," *Yale Global* (October 1, 2003), http://yaleglobal.yale.edu/display.article?id=2550 (accessed June 3, 2006); "China's Manned Space Program: Sun Tzu or Apollo Redux?" *Naval War College Review* 56 (Summer 2003), 51–71; and "'Houston, We Have a Problem': China and the Race to Space," *Current History* (September 2003), 259–265.
4 *Corona Fact Sheet*, National Reconnaissance Office, http://www.nro.gov/corona/facts.html (accessed April 12, 2006).
5 Theodore Shabad, "Communist China's Five Year Plan," *Far Eastern Survey* 24 (12) (December 1955), 189–191.
6 York W. Bradshaw, Young-Jeong Kim and Bruce London, "Transnational Economic Linkages, the State, and Dependent Development in South Korea, 1966–1988: A Time-Series Analysis," *Social Forces* 72 (2) (December 1993), 315–345; Eun Mee Kim, "Global Perspectives on Social Problems – Contradictions and Limits of a Developmental State: With Illustrations from the South Korean Case," *Social Problems* 40 (2) (May 1993), 228–249; Bae-Gyoon Park, "Where Do Tigers Sleep at Night? The State's Role in Housing Policy in South Korea and Singapore," *Economic Geography* 74 (3) (July 1998), 272–288.
7 William A. Joseph, "A Tragedy of Good Intentions: Post-Mao Views of the Great Leap Forward," *Modern China* 12 (4) (October 1986), 419–457.
8 T.J. Hughes, "China's Economy – Retrospect and Prospect," *International Affairs* (Royal Institute of International Affairs 1944–) 46 (1) (January 1970), 63–73.
9 Appu Kuttan Soman, *Double-Edged Sword: Nuclear Diplomacy in Unequal Conflicts – The United States and China, 1950–1958* (Westport, CN: Praeger, 2000).

10 Stephen Van Evera, *Causes of War: Power and the Roots of Conflict* (Ithaca, NY: Cornell University Press, 1999).
11 Allen S. Whiting, *China Crosses the Yalu: The Decision to Enter the Korean War* (Stanford, CA: Stanford University Press, 1980). Edward Friedman, Harold C. Hinton, Allen S. Whiting and Jerome Alan Cohen, *Taiwan and American Policy* (New York: Praeger, 1971).
12 Joseph A. Camilleri, *Chinese Foreign Policy: The Maoist Era and its Aftermath* (Seattle: University of Washington Press, 1980).
13 Dwight D. Eisenhower, *The White House Years* (Garden City, NY: Doubleday, 1963–1965).
14 Zalmay M. Khalilzad, *The United States and a Rising China*: Strategic and Military Implications (Santa Monica, CA: RAND, 1999).
15 Alexander L. George, *Deterrence in American Foreign Policy: Theory and Practice* (New York, Columbia University Press, 1974), 235.
16 *Ibid*.
17 David Shambaugh, "Nationalism and Internationalism: Sino-Soviet Relations," in Thomas W. Robinson and David Shambaugh (eds), *Chinese Foreign Policy: Theory and Practice* (Oxford: Clarendon Press; New York: Oxford University Press, 1994).
18 Qiang Zhai, *China and the Vietnam Wars, 1950–1975* (Chapel Hill: University of North Carolina Press, 2000).
19 Harlan W. Jencks, *From Muskets to Missiles: Politics and Professionalism in the Chinese Army, 1945–1981* (Denver, CO: Westview Press, 1982), 57.
20 Faren Qi, "Mao Zedong and Dong Fang Hong Satellite," *Aerospace China* 4 (2000).
21 Yiguang Qu, "Historical Review on the Research of Long March Series," *Aerospace China* 4 (1998), http://www.space.cetin.net.cn/docs/ht9804/ht980413.HTM
22 GGDC. www.ggdc.net/dseries/ (accessed 27 December, 2005).
23 William Burr and Jeffrey T. Richelson, "'Whether to Strangle the Baby in the Cradle': The United States and the Chinese Nuclear Program, 1960–1964," *International Security* (Winter, 2000–2001), 54–99.
24 Jiaqi Yan and Gao Gao, *Turbulent Decade: A History of the Cultural Revolution* (Honolulu: University of Hawai'i Press, 1996).
25 Joan Robinson, "The Cultural Revolution in China," *International Affairs* (Royal Institute of International Affairs) 44 (2) (April 1968), 214–227.
26 Lowell Dittmer, *Liu Shao-ch'i and the Chinese Cultural Revolution: The Politics of Mass Criticism* (Berkeley: University of California Press, 1974).
27 *The Great Socialist Cultural Revolution in China* (Peking: Foreign Languages Press, 1966), III, 11–17. Editorial of the Liberation Army Daily (*Jiefangjun Bao*): "Mao Tse-Tung's Thought is the Telescope and Microscope of Our Revolutionary Cause," June 7, 1966.
28 Zhou Enlai, "Mao Zedong Thought Is the Sole Criterion of Truth," in Michael Schoenhals (ed.), *China's Cultural Revolution, 1966–1969 Not a Dinner Party* (Armonk, NY: M.E. Sharpe, 1996), 27.
29 Harvey, *The Chinese Space Programme: From Conception to Future Capabilities*.
30 John Wilson Lewis and Hua Di, "China's Ballistic Missile Programs: Technologies, Strategies, Goals (in Weapons Diffusion)," *International Security* 17 (2) (Autumn 1992), 5–40.
31 Information Bank Abstract, *New York Times* (April 26, 1970), section 4, p. 12, col. 1.
32 Information Bank Abstract, *New York Times* (April 26, 1970), section 4, p. 3, col. 7.

33 Guoxiagn Wang, "Divine Arrow Fights from Here," *Areospace China* (October 1999).
34 Hughes, "China's Economy – Retrospect and Prospect."
35 Kuang Huan Fan, *Mao Tse-tung and Lin Piao*: Post-revolutionary Writings, ed. K. Fan (Garden City, NY: Anchor Books, 1972).
36 *Ibid*.
37 *Ibid*.
38 Roderick Macfarquhar, "The Succession to Mao and the End of Maoism," in Denis Twitchett and John K. Fairbank (eds), *The Cambridge History of China* (Cambridge and New York: Cambridge University Press, 1978).
39 John Wilson Lewis (ed.) *Party Leadership and Revolutionary Power in China* (Cambridge: Cambridge University Press, 1970), 81.
40 John Gittings, *The Role of the Chinese Army* (New York: Oxford University Press, 1967), 72.
41 Macfarquhar, "The Succession to Mao and the End of Maoism."
42 Macfarquhar, "The Succession to Mao and the End of Maoism."
43 Iris Chang, *Thread of the Silkworm* (New York: Basic Books, 1995).
44 Jiang Huang, *Factionalism in Chinese Communist Politics* (Cambridge and New York: Cambridge University Press, 2000), 74.
45 Harvey, *The Chinese Space Programme*.
46 *Ibid*.
47 *Ibid*.
48 Zhen Li and Roger Handberg, "The Central Dilemma of China's S&T Policy," *Bulletin of Science, Technology & Society* 22 (2002), 484–495.

4 Accelerating the rise of China's space program

1 Roger Handberg, "Outer Space as a Shared Frontier: Canada and the United States, Cooperation between Unequal Partners," *American Behavioral Scientist* 47 (June 2004), 1263–1277.
2 R. Hu, "Chinese Space and Aviation Industries Score Major Breakthroughs," (Washington, DC: National Aeronautics and Space Administration, April 1986, NAS 1.1587973, NASATM-87973), 16p.
3 C.N. Hill, *A Vertical Empire: The History of the UK Rocket and Space Program. 1950–1971* (London: Imperial College Press, 2001).
4 Deng Xiaoping, "Use International Intelligence and Broaden the Scope of Opening Up," *Selected Work of Deng Xiaoping, Volume III* (Beijing: Renmin Press, 1993), 32.
5 Joseph Y.S. Cheng, *China: Modernization in the 1980s* (Hong Kong: Chinese University Press; reprinted New York: St. Martin's Press, 1990).
6 Nicholas R. Lardy, "Recasting the Economic System," in Michael Ying-Mao Kau and Susan H. Marsh (eds), *China in the Era of Deng Xiaoping* (Armonk, NY: M.E. Sharpe, 1993), 106.
7 Kungchia Yeh, "Foreign Trade, Capital Inflow and Technology Transfer," in Michael Ying-Mao Kau and Susan H. Marsh (eds), *China in the Era of Deng Xiaoping* (Armonk, NY: M.E. Sharpe, 1993).
8 David S. Goldman, *Deng Xiaoping and the Chinese Revolution: A Political Biography* (New York: Routledge, 1995).
9 Liu Jiyuan, "Space for Development," *Harvard International Review* 16 (Summer 1994), 36.
10 "Report: Japan to Launch Two Spy Satellites by March 2007," *Space.com* (January 6, 2006) http://www.space.com/missionlaunches/ap_060106_japan_spysats.html (accessed May 21, 2006).

11 Shen Liu, "The Development of Satellite Communications of China," *Aerospace China* (English edition), 1 (Winter 1992), 7, 8; Guonui Chen and Warren Stutzman, "Satellite Communications in China," *Space Communications* 9 (December 1991), 9–14.
12 "Indonesian Rebels End 29-Year Insurgency," *Chinadaily.com.cn* (December 28, 2005), http://www.chinadaily.com.cn/english/doc/2005-12/28/content_507308.htm (accessed May 29, 2006).
13 Weixing Jiang, "The Development and the Future Applications of the Chinese Oceanic Satellites," *Aerospace China* 9 (2001).
14 Jiyuan Liu, "The Application of Chinese Space Technology and Its Benefit," *Aerospace China* 11 (2001).
15 Ibid.
16 Roger Handberg, *International Space Commerce: Building from Scratch* (Gainesville, FL: University Press of Florida, 2006), ch. 8.
17 Ibid., ch. 3.
18 Liheng Wang, "Seize Today and Look Forward to Tomorrow," *Aerospace China* (1002–7742) 186, (October 1993), 4–6.
19 James C. Mulvenon and Andrew N.D. Yang. *Seeking Truth From Facts: A Retrospective on Chinese Military Studies in the Post-Mao Era* (Santa Monica, CA: National Security Research Division, RAND, 2001), 97.
20 Deng Xiaoping. *Selected Works of Deng Xiaoping. People's Daily*, http://english.peopledaily.com.cn/dengxp/vol3/text/c1570.html (accessed May 28, 2002).
21 Cheng, *China: Modernization in the 1980s*.
22 Yanping Chen, "China's Space Policy – a Historical Review," *Space Policy* (May 1991), 116–128.
23 The contrast between Korolov as Chief Designer and his successors was dramatic, see Asif A. Siddiqi, *Sputnik and the Soviet Space Challenge* (Gainesville, FL: University Press of Florida, 2003; earlier published as *Challenge to Apollo: The Soviet Union and the Space Race: 1945–1974*, National Aeronautics and Space Administration, History Division, SP-2000-4408).
24 Brian Harvey, *Race into Space: The Soviet Space Program* (Chichester, West Sussex: E. Harwood; reprinted New York: Halsted Press, 1988).
25 Deng Xiaoping, "Science and Technology is the Primary Productive Force," *Selected Works of Deng Xiaoping. Volume III* (Beijing: Renmin Press, 1993).
26 Deng Xiaoping, "Priority Should Be Given to R&D," *Selected Works of Deng Xiaoping. Volume II* (Beijing: Renmin Press, 1994), 32.
27 Deng Xiaoping, "The Speech on the National Science Conference," *Selected Works of Deng Xiaoping. Volume II* (Beijing: Renmin Press, 1994), 91.
28 Deng Xiaoping, "China Must Take Its Place in the Field of High Technology," *Selected Works of Deng Xiaoping. People's Daily*, http://english.people.com.cn/dengxp/vol3/text/c1920.html (accessed May 21, 2006).
29 Roger Handberg, *Ballistic Missile Defense and the Future of American Security* (Westport, CN: Praeger, 2002), 61–82.
30 "Deploying US Missile Defences: Technical Problems, Policy Questions," *Strategic Comments* 9 (1) (January 2003), http://www.iiss.org/publications/strategic-comments/past-issues/volume-9---2003/volume-9---issue-1/deploying-us-missile-defences (accessed May 29, 2006).
31 Huabao Lin, "China's New Recoverable Satellite Recovery System," 13th Aerodynamic Decelerator Systems Technology Conference, Clearwater Beach, FL, May 15–18, 1995, Technical Papers (A95-30501 07-03) (Washington, DC: American Institute of Aeronautics and Astronautics, 1995), 127–129; Harvey, *Race into Space*.

32 Lou Ge, "China's Space Program," Center for Security and International Studies (April 3, 2006), http://www.csis.org/media/csis/events/060403_china_space.pdf (accessed April 11, 2006).
33 "Launch Log," Go Taikonauts?, http://www.geocities.com/CapeCanaveral/Launchpad/1921/launch.htm (accessed May, 2005).
34 Wayne Eleazer, "First Flight Facts," *The Space Review* (May 8, 2006), http://www.thespacereview.com/article/616/1 (accessed May 9, 2006).
35 The United States House of Representatives Select Committee on US National Security and Military/Commercial Concerns with the People's Republic of China, "The Cox Report" (January 9, 1999), http://www.house.gov/coxreport/cont/gncont.html (accessed May 21, 2006).
36 *Strom Thurmond Defense Act Defense Authorization Act for Fiscal Year 1999*, PL 105–261, 105th Congress (October 18, 1998).
37 *International Traffic in Arms Regulations* (ITAR) (22 CFR 120–130).
38 Jiyuan Liu, "Strengthening Sino-Germans pace Cooperation Looking Forward to 21st Century," *China Engineering Technology and Space Information*, http://www.space.cetin.net.cn/docs/HTM-E/002.htm (accessed May 21, 2006).
39 Information Office of the State Council of the People's Republic of China. "China's National Defense, 2004," http://news.xinhuanet.com/mil/2004-12/27/content_2384731_4.htm (accessed May 22, 2006).
40 PBS, "Arming Taiwan," http://www.pbs.org/newshour/bb/asia/jan-june01/taiwan_arms.html (accessed May 21, 2006).
41 John Pomfret, "China to Buy 8 More Russian Submarines," Washington Post Foreign Service, *Washington Post* (June 25, 2002), A15.
42 Department of Defense of the United States, *Quadrennial Defense Review Report* (Washington, DC: Office of the Secretary of Defense, September 30, 2001) (accessed October 2, 2001).
43 Handberg, *Ballistic Missile Defense*.
44 Avery Goldstein, "The Diplomatic Face of China's Grand Strategy: A Rising Power's Emerging Choice," *The China Quarterly* (December 2001), 835.
45 Victoria Samson, "An 'F' for Missile Defense: How Seven Government Reports in Two Months Illustrate the Need for Missile Defense to Change its Ways" (Washington: Center for Defense Information, April 25, 2006).
46 Tai Ming Cheung, "Reforming the Dragon's Tail: Chinese Military Logistics in the Era of High-Technology Warfare and Market Economics," in James R. Lilley and David Shambaugh (eds), *China's Military Faces the Future* (Washington, DC: AEI; Armonk, NY: M.E. Sharpe, 1999).
47 *Ibid.*
48 "Keep an Eye on the Military Cooperation Between India and the US," *Liberation Daily* (February 24, 2002).
49 *1998 National Defense Report*, Republic of China (Taiwan, 1998).
50 Alasdair McLean, "A New Era? Military Space Policy Enters the Mainstream," *Space Policy* 16 (November 2000), 224–245.
51 *Ibid.*, 237–239.
52 Cheung, "Reforming the Dragon's Tail 70.
53 You Ji, *The Armed Forces of China* (London and New York: I.B. Tauris, 1999), 83.
54 *Ibid.*
55 "ZY," *Encyclopedia Astronautica*, http://www.astronautix.com/craft/zy.htm (accessed June 2, 2002).
56 Ho Cheng, "China Eyes Anti-Satellite System," *Space Daily* (January 8, 2000), http://www.spacedaily.com/news/china-01c.html (accessed March 26, 2002).

57 "China Eyes US Military in S. Pacific," *Christian Science Monitor* (October 30, 1997), 7A.
58 Cheng, "China Eyes Anti-Satellite System."
59 Gregory Kulacki and David Wright, "A Military Intelligence Failure? The Case of the Parasite Satellite," Union of Concerned Scientists (August 16, 2004), http://www.ucsusa.org/global_security/china/page.cfm?pageID=1479 (accessed May 21, 2006).
60 "China Starts Research Plans on Space Debris," *Aerospace China* 6 (2001).
61 US Defense Department, *Annual Report to Congress on the Military Power of the People's Republic of China* (Washington, DC: Government Printing Office, 2006), http://permanent.access.gpo.gov/LPS24358/2000_report.pdf (accessed May 21, 2006).
62 David B. Sandalow, "Remote Sensing and Foreign Policy," Symposium on "Viewing the Earth: The Role of Satellite Earth Observations and Global Monitoring in International Affairs" (Washington, DC, George Washington University, June 6, 2000).
63 John Baker, "Mitigating the South China Sea Disputes Through Cooperative Monitoring," Symposium on "Viewing the Earth: The Role of Satellite Earth Observations and Global Monitoring in International Affairs" (Washington, DC, George Washington University, June 6, 2000).
64 Jiang, "The Development and the Future Applications of the Chinese Oceanic Satellites."
65 Vernon Van Dyke, *Pride and Power: The Rationale of the Space Program* (Urbana: University of Illinois Press, 1964), 135.
66 "Decoding the Mystery of China's Manned Space Program," *Liberation Daily*, http://old.jfdaily.com/gb/node2/node172/node20519/node20523/userobject1ai304563.html (accessed May 22, 2006).
67 Gene Krantz, *Failure Is Not an Option: Mission Control From Mercury to Apollo 13 and Beyond* (New York: Simon & Schuster, 2000).
68 *The Independent* (London) (November 22, 1999).
69 Michael Laris, "Chinese Test Craft For Manned Orbits, Space Launch Boosts National Pride," *Washington Post* (November 22, 1999), A01.
70 James E. Webb, Address of June 7, 1961, NASA News Release No. 61-124, p. 7
71 Craig Covault, "China Seeks ISS Role, Accelerates Space Program," *Aviation Week & Space Technology* (November 12, 2001), 52.
72 Richard Nixon, *Statement by President Nixon on the Space Program*, March 7, 1970. Released from the Office of the White House Press Secretary, Key Biscayne, Florida.
73 Dennis Normile with Ding Yimin, "Science Emerges from Shadows of China's Space Program," *Science* 296 (June 7, 2002), 1788–1791.
74 "China to Build World's Largest Satellite System," *People's Daily* (September 21, 2001).
75 Fangying Cou, "Hold the Chance to Develop Satellite Communication," *Science and Technology Daily* (China) (June 1, 1999).
76 Bangzhi Tan, *Aerospace China* 7 (1998); Lehao Long, "China Report," *Aerospace China* 3 (2001).
77 Zhu Rongji, "Report on the Outline of the Tenth Five-Year Plan for National Economic and Social Development," http://www.china.org.cn/ch-15/plan9.htm (accessed May 22, 2006).
78 "PRC BJB Discusses Planned Space Exploration," *Liberation Daily* (December 3, 2001).

79 Dengqui Lu, "Current Application and Vista of Space Technology Used in Farmland Management," *Aerospace China* 7 (1998).
80 Qunfang Hu and Guoying Chen, "China is to Develop Satellites in Monitoring Its Environment and Forecasting Disaster," *China Aerospace* News (September 27, 2003).
81 Peter Creola, "A Long-term Space Policy for Europe," *Space Policy* 15 (1999), 207–211.
82 Rongji, "Report on the Outline of the Tenth Five Year Plan for National Economic and Social Development."
83 Zhen Li and Roger Handberg, "The Central Dilemma of China's S&T Policy," *Bulletin of Science, Technology & Society* 22 (2002), 484–495.
84 *Ibid.*
85 Central Intelligenc Agency, "China," http://www.cia.gov/cia/publications/factbook/geos/ch.html (accessed May 22, 2004).

5 The politics of Chinese human spaceflight

1 Anatoly Zak, "Kliper: Key Elements," http://www.russianspaceweb.com/Kliper.html (accessed March 15, 2006).
2 Brian Harvey, *China's Space Program: From Conception to Manned Spaceflight* (New York: Springer-Praxis, 2004), 24–25.
3 Martin Sieff, "Analysis: China Threatens Japan's UN Hopes," *UPI* (April 11, 2005), http://www.spacewar.com/upi/2005/WWN-UPI-20050411-21001800-bc-china-japan-anal (accessed April 13, 2005).
4 E.P. Grondine, "Chinese Manned Space Program: Behind Closed Doors," *Encyclopedia Astronautica*, http://www.astronautix.com/articles/chidoors.htm (accessed January 10, 2005).
5 David J. Shayler, *Disasters and Accidents in Manned Spaceflight* (New York: Springer, Praxis, 2000).
6 Dwayne A. Day, "The Development and Improvement of the CORONA Satellite," in Dwayne A. Day, John M. Logsdon and Brian Latell (eds), *Eye in the Sky: The Story of the Corona Spy Satellites* (Washington: Smithsonian Institution Press, 1998).
7 Michael Coren, "Genesis Capsule Crashes in the Desert," *CNN.com* (September 8, 2004), http://www.cnn.com.2004/TECH/space/09/08/genesis.entry.com (accessed March 26, 2005).
8 *Ibid.*
9 Joan Johnson-Freese, "China's Manned Space Program: Sun Tzu or Apollo Redux?" *Naval War College Review* 56 (Summer 2003), 51–71.
10 Jean-Jacques Servan-Schreiber, *The American Challenge*, trans. Robert Steel (New York: Avon Books, 1969).
11 The discussion here draws on Harvey, *China's Space Program*; and "Shuguang-1," *Encyclopedia Astronautica*, http://www.astronautica.com/craft/shuuang1.htm (accessed May 24, 2005).
12 All three terms have been used to describe Chinese flight personnel; the latter two have been more common, with taikonauts more prevalent in English. Generally, taikonauts will be used here.
13 Qing Zhao. "Unveil the Mystery of the First Chinese Astronauts Selection in the 1970s," *People's Daily*, http://www.people.com.cn/GB/keji/1059/2137459.html (accessed May 23, 2006).
14 "Chinese Manned Capsule 1978," *Encyclopedia Astronautica*, http://www.astronautix.com/craft/chic1978.htm (accessed May 24, 2005); Grondine, "Chinese Manned Space Program: Behind Closed Doors."

15 Dean Cheng, "Dragon in Orbit: A Survey of the Chinese Space Program," Twenty-Fifth Annual International Space Development Conference (May 2005), 2.
16 Chen Lan, "Shenzhou History," http://www.geocities.com/CapeCanaveral/Launchpad/1921/story-7.htm?200517 (accessed May 17, 2005).
17 Harvey, *China's Space Program*, 247; "Project 921," *Encyclopedia Astronautica*, http://www.astronautix.com/articles/china.htm (accessed May 25, 2005).
18 Mark Wade, "Shenzhou," *Encyclopedia Astronautia* (January 6, 2006), http://www.astronautrix.com/craft/shenzhou.htm (accessed February 15, 2006).
19 "Space Craft: Manned, Soyuz," http://www.russianspaceweb.com/soyuz.html (accessed 14 February, 2006).
20 "AIP FYI #138: NASA Administrator Griffin and Congress: NASA's Exploration Architecture," *NASA Watch* (September 23, 2005), http://www.spaceref.com/news/viewsr.html?pid=18155 (accessed October 3, 2005).
21 Mark Wade, "Shenzhou," *Encyclopedia Astronautia* (January 6, 2006), http://www.astronautrix.com/craft/shenzhou.htm (accessed February 15, 2006).
22 Asif A. Siddiqi, *Sputnik and the Soviet Space Challenge* (Gainesville, FL: University Press of Florida, 2003; earlier published as *Challenge to Apollo: The Soviet Union and the Space Race: 1945–1974*, National Aeronautics and Space Administration, History Division, SP-2000-4408), ch. 6.
23 Constance McLaughlin Green and Milton Lomask, *Vanguard: A History* (Washington, DC: Smithsonian Institution Press, 1971).
24 Wei Long, "Shenzhou-2 Returns While Orbital Experiments Continue," *Space Daily* (January 19, 2001), http://www.spacedaily.com/news/china-01m.html (accessed January 30, 2001).
25 Wei Long, "Shenzhou-3 Orbit Module Continues Experiments," *Space Daily* (June 28, 2001), http://www.spacedaily.com/news/china-02zp.html (accessed July 9, 2001).
26 Leonard David, "China's Shenzhou 4 Set for Weekend Homecoming," *Space.com* (January 3, 2003), http://www.space.com/missionlaunches/shenzhou_update_030103.html (accessed January 9, 2003).
27 Joan Johnson-Freese, "Testimony Before the US–China Economic and Security Review Commission: China Military Modernization and Cross-Strait Balance" (September 15, 2005); Joan Johnson-Freese, "China Bids for the High Ground," *Yale Global Online* (October 1, 2003), http://yaleglobal.yale.edu/article.print?=2550 (accessed April 8, 2006).
28 Jim Banke, "China Launches Its First Piloted Space Flight," *Space.com* (October 15, 2003), http://www.space.com/missionlaunches/shenzhou5_launch_031014.html (accessed October 16, 2003).
29 Morris Jones, "Fly Me to a Red Moon," *SpaceDaily.com* (March 28, 2006), http://www.spacedaily.com/reports/Fly_Me_To_A_Red_Moon.html (accessed March 30, 2006).
30 "Spacecraft: Manned: Salyut Era," http://www.russianspaceweb.com/spacecraft_manned_salyut.html (accessed 21 January, 2006).
31 Roger Handberg, "The Fluidity of Presidential Policy Choice: The Space Station. The Russian Card and US Foreign Policy," *Technology in Society* 20 (1998), 421–439.
32 Dwayne A. Day, "Red Moon, Dark Moon," *The Space Review* (October 11, 2005), http://www.thespacereview.com/article/473/1 (accessed October 11, 2005).
33 "Experts on China's Lunar Probing Program," *China.org.cn* (March 23, 2004), http://www.china.org.cn/english/scitech/91021.htm (accessed May 19, 2006). This is a collection of three essays written for *People's Daily*.

34 Leonard David, "China's Lunar Ambitions Have Produced Tensions at Home," *Space.com* (October 3, 2003), http://www.space.com/missionlaunches/china_lunar_031007.html (accessed October 10, 2003).

6 Assessing China's future in space

1 Office of the Secretary of Defense (OSD), *Annual Report to Congress: The Military Power of the People's Republic of China, 2005* (Washington: Office of Secretary of Defense, 2005), 14.
2 Office of the Secretary of Defense (OSD), *Annual Report to Congress: The Military Power of the People's Republic of China, 2006* (Washington: Office of Secretary of Defense, 2006), 31–36.
3 Richard P. Suttmeier, Cong Cao and Denis Fred Simon, "'Knowledge Innovation' and the Chinese Academy of Sciences," *Science* 312 (April 7, 2006), 58–59; Hao Xin and Gong Yidong, "China Bets Big on Big Science," *Science* 312 (March 17, 2006), 1548–1549; Ling Zu, "China Issues S&T Development Guidelines," *Xinhua* (February 9, 2006), http://english.gove.cn/2006-02/09/content_183426.htm (accessed April 1, 2006).
4 Roger Handberg, *International Space Commerce* (Gainesville, FL: University Press of Florida, in press), ch. 2.
5 David E. Lupton, *On Space Warfare: A Space Power Doctrine* (Maxwell AFB, AL: Air University Press, June 1988).
6 Lorenza Sebesta, *United States–European Cooperation in Space during the Sixties* (Noordwijik, Netherlands: ESA Publications Division, July 1994); John Krige, *The Launch of ELDO* (March 1993).
7 "Nigerian Satellite to be Launched from Xichang Launch Center," *Spacedaily* (April 18, 2005), http://www.spacedaily.com/news/china-05zn.html (accessed April 20, 2005).
8 Peter B. de Selding, "EU and China Collaborate on Galileo Navigation Satellite Project," *Space.com* (September 9, 2003), http://www.space.com/news/china_galileo_030921.html (accessed October 10, 2003); "First Orbiting Galileo Navigation Satellite Speeds Up China–Europe Space Cooperation," *People's Daily Online* (December 29, 2005), http://english.people.com.cn/200512/29/eng20051229_23158.html (accessed May 7, 2006).
9 "European Global Position Satellites: Galileo Navigation Satellites," *Space Today Online*, http://222.spacetoday.org/Satellites/GalileoEuroNavSat.html (accessed May 15, 2006).
10 Zhang Hui, "Security in Space," *China Security* (Washington, DC: World Security Institute China Program, 2006), 29.
11 Roger Handberg, *Seeking New World Vistas: The Militarization of Space* (Westport, CN: Praeger, 2000), 57–58; Paul B. Stares, *The Militarization of Space: US Policy: 1945–1984* (Ithaca, NY: Cornell University Press, 1985).
12 *Convention on International Liability for Damage Caused by Space Objects* (Resolution 2777 [XXII] annex) (November 29, 1971).
13 Bruce DeBlois, "Space Sanctuary: A Viable National Strategy," *Aerospace Power Journal* (Winter 1988), 41–57.
14 Handberg, *Seeking New World Vistas*, 88–90.
15 OSD, *Annual Report 2005*, 36.
16 Gregory Kulacki and David Wright, *A Military Intelligence Failure? The Case of the Parasite Satellite* (Cambridge: Union of Concerned Scientists, August 14, 2004), http://www.ucsusa.org/global_security/china/a-military-intelligence-failure-the-chinese-parasite-satellite.html (accessed March 7, 2006).

17 M.V. Smith, *Ten Propositions Regarding Spacepower* (Maxwell AFB, AL: Air University Press, October 2002); Simon P. Worden and John E. Shaw, *Whither Space Power? Forging a Strategy for the New Century* (Maxwell AFB, AL: Air University Press, September 2002).
18 Samuel P. Huntington, *The Common Defense* (New York: Columbia University Press, 1961).
19 John Lewis Gaddis, *The Cold War: A New History* ((New York: Penguin, 2005).
20 Joan J. Johnson-Freese and Roger Handberg, "Deja Vu All Over Again? The Myth of a National Space Transportation Policy," *Defense Analysis* 12 (1996), 161–172.
21 *US Commercial Remote Sensing Policy* (Washington: White House, Office of Science and Technology Policy, April 25, 2003).
22 State Council of the People's Republic of China, *White Paper: China's Space Activities* (Beijing, China: Xinhua News Agency, August 6, 2004). Online at: http:www.spacedaily.com/news/China-04zw.html (accessed March 15, 2005).
23 William Burr and Jeffrey T. Richelson, "Whether to 'Strangle the Baby in the Cradle': The United States and the Chinese Nuclear Program, 1960–1964," *International Security* 25 (Winter 2000), 54–99.
24 Asif A. Siddiqi, *The Soviet Space Race with Apollo* (Gainesville, FL: University Press of Florida, 2003).
25 Deborah Zabarenko, "Chinese Space Official Lays Out Exploration Plans," *Reuters* (April 4, 2006), http://today.reuters.com/misc/PrinterFriendlyPopup.aspx?type=scienceNews&storyID=ciri:2 . . . (accessed April 4, 2006).
26 Iridium, http://www.astronautix.com/project/iridium.htm (accessed May 21, 2005).
27 State Council, *White Paper*, 2.
28 Lon Rains, "NASA Chief Michael Griffin Invited to China," *Space.com* (April 5, 2006), http://www.space.com/news/060405_nss_griffin.html (accessed April 6, 2006).
29 Keith Cowing, "China Will Propose that NASA Use Shenzhou for ISS Duty" (March 1, 2005), http://www.spaceref.com/news/viewnews.html?id=1004 (accessed April 5, 2006).
30 Frank Morring, Jr., "'Global Security,'" *Aviation Week & Space Technology* (May 1, 2006), 26.
31 Craig Covault, "The China Card: US Now Agreeable to Space Cooperation with China," *Aviation Week & Space Technology* (January 31, 2005), 27.
32 Vincent G. Sabathier, "Europe and China," *Ad Astra* (May 5, 2005), http://www.space.com/adastra/china_europe_0505.html (accessed May 26, 2005).
33 Roger Handberg, "Outer Space as a Shared Frontier: Canada and the United States, Cooperation between Unequal Partners," *American Behavioral Scientist* 47 (June 2004), 1263–1277.
34 Joan Lisa Bromberg, *NASA and the Space Industry* (Baltimore, MD: Johns Hopkins University Press, 2000).
35 Steven Berner, *Japanese Space Program: A Fork in the Road?* (Santa Monica, CA: RAND, 2005).
36 Larry Wheeler, "US Losing Space Race, Congressmen Say," *Floridatoday.com* (March 31, 2006), http://www.floridatoday.com/apps/pbcs.d11/article?AID=/20060331 (accessed April 3, 2006).
37 "China's Moon Quest Has US Lawmakers Seeking New Space Race," *Bloomberg.com* (April 20, 2006), http://www.bloomberg.com/opps/news?pid=7100000/refer=asia&sid=a5elzidanccM (accessed April 22, 2006).

38 Mark Carreau, "Hutchison: US Should Attempt to Finish Building Space Station," *Houston Chronicle* (December 8, 2005), http://www.chron.com/cs/cda/printstory.mpl/space/3510547 (accessed January 7, 2006).
39 Roger Handberg, *Reinventing NASA* (Westport, CN: Praeger, 2003), appendix.
40 "Asia-Pacific Countries Sign Space Cooperation Convention," *People's Daily Online* (October 28, 2005), http://english.people.com.cn/200510/28/print20051028_217581.html (accessed May 20, 2006); Leonard David, "China-Based Asia Pacific Space Group Gains Support," *Space.com* (November 12, 2003), http://www.space.com/apsco_support_031112.html (accessed May 13, 2006).
41 "Gathering for Proposed Asia-Pacific Space Cooperation Organization," *Chinadaily.com.cn* (November 10, 2003), http://www.chinadaily.com.cn/doc/2003-11/10/content_280249.htm (accessed May 14, 2006).
42 "Bush Offers China Space Cooperation," *The Sydney Morning Herald* (April 22, 2006), http://www.smh.com.au/news/World/Bush-offers-China-space-cooperation/2006/04/22/11 (accessed April 23, 2006).
43 Leonard David, "US–China Space Ties Weighed," *Space.com* (April 20, 2006), http://www.space.com/news/060420-China-visit.html (accessed April 22, 2006).
44 Morring, "'Global Security,'" 26.
45 Handberg, *Reinventing NASA*, ch. 4.
46 Roy Gibson, "China's Long March into the Space Age," *Nature* 393 (6682) (May 21, 1998), 225.
47 George Desvaux, Michael Wang and David Xu, "Spurring Performance in China's State Owned Enterprises," *McKinsey Quarterly* (March 2006), http://www.mckinseyquarterly.com/article-page.aspx?ar=1492&L2-19&13=69&srid=7&... (accessed May 25, 2006); "Initial Success Achieved in State Owned Enterprise Economic Restructuring," *People Daily Online* (November 16, 2001), http://english.people.com.cn/200411/16/print20041116_164083.html (accessed May 25, 2006).
48 Theresa Hitchens, *Developments in Military Space: Movement toward Space Weapons* (October 2003), http://www.cdi.org/pdfs/space-weapons.pdf (accessed May 25, 2006); Gerald M. Steinberg, *Satellite Capabilities of Emerging Space Competent States*, http://faculty.biu.ac.il/~steing/military/sat.htm (accessed May 25, 2006); Jerrold F. Elkin and Brian Fredericks, "Military Implications of India's Space Program," *Air University Review* (May–June 1983), http://www.airpower.maxwell.af.mil/airchronicles/aureview/1983/May-Jun/fredericks.htm (accessed April 7, 2005).

Selected references

The references listed below are the major references used in the work; other works cited are listed in the notes for the individual chapters. Those in the notes cover more specific events or issues of interest. The citations listed are a combination of relevant books, government and other reports, and articles both in print and from the web.

Books and reports

Allison, Graham T. 1971. *Essence of Decision: Explaining the Cuban Missile Crisis.* Boston: Little, Brown.
Berner, Steven. 2005. *Japanese Space Program: A Fork in the Road?* Santa Monica, CA: RAND.
Bromberg, Joan Lisa. 2000. *NASA and the Space Industry.* Baltimore, MD: Johns Hopkins University Press.
Bulkeley, Rip. 1991. *The Sputniks Crisis and Early United States Space Policy.* Bloomington, IN: Indiana University Press.
Burrows, William E. 1998. *This New Ocean: The Story of the First Space Age.* New York: Modern Library.
Camilleri, Joseph A. 1980. *Chinese Foreign Policy: The Maoist Era and its Aftermath.* Seattle: University of Washington Press.
Central Intelligence Agency. 2004. "China." Washington: CIA. http://www.cia.gov/cia/publications/factbook/geos/ch.html (accessed May 22, 2004).
Chairman of the Joint Chiefs of Staff. 2004. *National Military Strategy of the United States of America* Washington: Office of the Joint Chiefs of Staff.
Chang, Iris. 1995. *Thread of the Silkworm.* New York: Basic Books.
Cheng, Joseph Y.S. 1990. *China: Modernization in the 1980s.* Hong Kong: Chinese University Press; New York: St. Martin's Press.
Convention on International Liability for Damage Caused by Space Objects (Resolution 2777 [XXII] annex), November 29, 1971.
De Maria, Michelangelo. 1993. *The History of the ELDO, Part 1: 1961–1964.* Noordwijk, Netherlands: European Space Agency, HSR-10.
Deng, Xiaoping. 1988. "China Must Take Its Place in the Field of High Technology," *Selected Work of Deng Xiaoping. People's Daily,* http://english.people.com.cn/dengxp/vol3/text/c1920.html (accessed May 21, 2002).

Deng, Xiaoping. 1985. "Priority Should Be Given to R&D," *Selected Work of Deng Xiaoping*. Volume II. Beijing: Renmin Press, 1994.

Deng, Xiaoping. 1988. "Science and Technology is the Primary Productive Force," *Selected Work of Deng Xiaoping*. Volume III. Beijing: Renmin Press, 1993.

Deng, Xiaoping. 1986. *Selected Work of Deng Xiaoping*. *People's Daily*, http://english.peopledaily.com.cn/dengxp/vol3/text/c1570.html (accessed May 28, 2002).

Deng, Xiaoping. 1985. "The Speech on the National Science Conference," *Selected Work of Deng Xiaoping*. Volume II. Beijing: Renmin Press, 1994.

Deng, Xiaoping. 1983. "Use International Intelligence and Broaden the Scope of Opening-up," *Selected Work of Deng Xiaoping*. Volume III. Beijing: Renmin Press, 1993.

Department of Defense of the United States. 2001. *Quadrennial Defense Review* (September 30, 2001) (accessed October 2, 2001).

Department of Defense of the United States. 2006. *Quadrennial Defense Review*. February 6, 2006.

Dittmer, Lowell. 1974. *Liu Shao-ch'i and the Chinese Cultural Revolution: The Politics of Mass Criticism*. Berkeley: University of California Press.

Divine, Robert A. 1993. *The Sputnik Challenge: Eisenhower's Response to the Soviet Satellite*. New York: Oxford University Press.

Dyke, Van. 1964. *Pride and Power; the Rationale of the Space Program*. Urbana: University of Illinois Press.

Friedman, Edward, Harold C. Hinton, Allen S. Whiting and Jerome Alan Cohen. 1971. *Taiwan and American Policy*. New York: Praeger.

George, Alexander L. 1974. *Deterrence in American Foreign Policy: Theory and Practice*. New York: Columbia University Press.

Gittings, John. 1967. *The Role of the Chinese Army*. New York: Oxford University Press.

Goldman, David S. 1995. *Deng Xiaoping and the Chinese Revolution: A Political Biography*. London: Routledge.

Handberg, Roger. 2002. *Ballistic Missile Defense and the Future of American Security*. Westport, CN: Praeger.

Handberg, Roger. 2006. *International Space Commerce: Building from Scratch*. Gainesville, FL: University Press of Florida.

Handberg, Roger. 2003. *Reinventing NASA*. Westport, CN: Praeger.

Handberg, Roger. 2000. *Seeking New World Vistas: The Militarization of Space*. Westport, CN: Praeger.

Handberg, Roger and Joan Johnson-Freese. 1994. *The Prestige Trap: A Comparative Study of the United States, European, and Japanese Space Programs*. Dubuque, IA: Kendall/Hunt Publishing.

Harvey, Brian. 2004. *China's Space Program: From Conception to Manned Spaceflight*. New York: Springer-Praxis.

Harvey, Brian. 1998. *The Chinese Space Program: From Conception to Future Capabilities*. Chichester and New York: Wiley.

Harvey, Brian. 1988. *Race into Space: The Soviet Space Program*. Chichester: E. Horwood; New York: Halsted Press.

Hill, C.N.A. 2001. *Vertical Empire: The History of the UK Rocket and Space Program, 1950–1971*. London: Imperial College Press.

Huang, Jiang. 2000. *Factionalism in Chinese Communist Politics.* Cambridge and New York: Cambridge University Press.
Information Office of the State Council of the People's Republic of China. 2004. *China's National Defense, 2004.* http://news.xinhuanet.com/mil/2004-12/27/content_2384731_4.htm (accessed May 22, 2006).
International Traffic in Arms Regulations (ITAR) (22 CFR 120–130).
Jencks, Harlan W. 1982. *From Muskets to Missiles: Politics and Professionalism in the Chinese Army, 1945–1981.* Denver, CO: Westview Press.
Johnson, Stephen B. 2002. *The Secret of Apollo: Systems Management in American and European Space Programs.* Baltimore, MD: Johns Hopkins University Press.
Johnson-Freese, Joan. 1998. *The Chinese Space Program: A Mystery Within a Maze.* Maibar, FL: Krieger Publishing Company.
Khalilzad, Zalmay M. 1999. *The United States and a Rising China: Strategic and Military Implications.* Santa Monica, CA: RAND.
Krige, John. 1993. *The Launch of ELDO.* Noordwijik, Netherlands: ESA Publications.
Lewis, John Wilson. ed. 1970. *Party Leadership and Revolutionary Power in China.* Cambridge: Cambridge University Press.
Lupton, David E. 1988. *On Space Warfare: A Space Power Doctrine.* Maxwell AFB, AL: Air University Press.
McDougall, Walter. 1985. . . . *The Heavens and the Earth: A Political History of the Space Age.* Baltimore, MD: Johns Hopkins University Press.
Office of the Secretary of Defense (OSD). 2006. *Annual Report to Congress: The Military Power of the People's Republic of China, 2006.* Washington: Office of Secretary of Defense (various years, same title used, most recent indicated).
Sebesta, Lorenza. 1996. *The Availability of American Launchers and Europe's Decision "To Go It Alone."* Noordwijk, Netherlands: European Space Agency, HSR-18.
Sebesta, Lorenza. 1993. *United States–European Cooperation in Space during the Sixties.* Noordwijik, Netherlands: ESA Publications.
Servan-Schreiber, Jean-Jacques. 1969. *The American Challenge,* trans. Robert Steel. New York: Avon Books.
Shayler, David J. 2000. *Disasters and Accidents in Manned Spaceflight.* New York: Springer, Praxis.
Short, Philip. 1999. *Mao: A Life.* New York: Henry Holt.
Siddiqi, Asif A. 2003. *The Soviet Space Race with Apollo.* Gainesville, FL: University Press of Florida.
Smith, M.V. 2002. *Ten Propositions Regarding Spacepower.* Maxwell AFB, AL: Air University Press.
Smith, Marcia S. 2004. *Potential International Cooperation in NASA's New Exploration Initiative.* Washington: Congressional Research Service, April 27.
Smith, Marcia. 2006. *Space Exploration: Issues concerning "Vision for Space Exploration."* Washington: Congressional Research Service, January 4.
Soman, Appu Kuttan. 2000. *Double-Edged Sword: Nuclear Diplomacy in Unequal Conflicts – The United States and China, 1950–1958.* Westport, CT: Praeger.
Stares, Paul B. 1985. *The Militarization of Space: US Policy: 1945–1984.* Ithaca, NY: Cornell University Press.

State Council of the People's Republic of China. 2000. *White Paper: 2000. China's Space Activities*. Beijing: Xinhua News Agency. Online at: http:www.spacedaily.com/news/China-04zw.html (accessed August 6, 2004).
Steinberg, Gerald M. *Satellite Capabilities of Emerging Space Competent States*. http://faculty.biu.ac.il/~steing/military/sat.htm (accessed May 25, 2006).
Stokes, Mark. 1999. *China's Strategic Modernization: Implications for the United States*. Carlisle, PA: Strategic Studies Institute.
United States House of Representatives Select Committee on US National Security and Military/Commercial Concerns with the People's Republic of China. 1999. *The Cox Report*. Washington: United States House of Representatives. http://www.house.gov/coxreport/cont/gncont.html (accessed May 21, 2006).
US Congress. 1998. *Strom Thurmond Defense Act Defense Authorization Act for Fiscal Year 1999*, PL 105–261 (105th Congress, October 18, 1998).
Van Evera, Stephen. 1999. *Causes of War: Power and the Roots of Conflict*. Ithaca, NY: Cornell University Press.
Whalen, David J. 2002. *The Origins of Satellite Communications, 1945–1965*. Washington, DC: Smithsonian Institution Press.
Whiting, Allen S. 1980. *China Crosses the Yalu: The Decision to Enter the Korean War*. Stanford, CA: Stanford University Press.
Worden, Simon P. and John E. Shaw. 2002. *Whither Space Power? Forging a Strategy for the New Century*. Maxwell AFB, AL: Air University Press.
Yan, Jiaqi and Gao Gao. 1996. *Turbulent decade: A History of the Cultural Revolution*. Honolulu: University of Hawai'i Press.
Zhai, Qiang. 2000. *China and the Vietnam Wars, 1950–1975*. Chapel Hill: University of North Carolina Press.

Articles and book chapters

Burr, William and Jeffrey T. Richelson. 2000. "Whether to 'Strangle the Baby in the Cradle': The United States and the Chinese Nuclear Program, 1960–1964," *International Security* 25 (Winter), 54–99.
Callahan, David and Fred I. Greenstein. 1997. "The Reluctant Racer: Eisenhower and US Space Policy," in Roger D. Launius and Howard E. McCurdy (eds), *Spaceflight and the Myth of Presidential Leadership*. Urbana: University of Illinois Press, 15–50.
Chen, Yanping. 1991. "China's Space Policy – a Historical Review," *Space Policy* (May), 116–128.
Cheng, Dean. 2005. "Dragon in Orbit: A Survey of the Chinese Space Program," Twenty-Fifth Annual International Space Development Conference (May).
Cheng, Ho. 2000. "China Eyes Anti-Satellite System," *Space Daily* (January 8), http://www.spacedaily.com/news/china-01c.html (accessed March 26, 2002).
Cheung, Tai Ming. 1999. "Reforming the Dragon's Tail: Chinese Military Logistics in the Era of High-Technology Warfare and Market Economics," in James R. Lilley and David Shambaugh (eds), *China's Military Faces the Future*. Washington, DC: AEI; Armonk, NY: M.E. Sharpe.
China.org.cn. 2004. "Experts on China's Lunar Probing Program," *China.org.cn* (March 23), http://www.china.org.cn/english/scitech/91021.htm (accessed May 19, 2006).

Christensen, Thomas J. 1999. "China, the US–Japan Alliance, and the Security Dilemma in East Asia," *International Security* 23 (Spring), 49–50.
DeBlois, Bruce. 1988. "Space Sanctuary: A Viable National Strategy," *Aerospace Power Journal* (Winter), 41–57.
Encyclopedia Astronautica. "Chinese Manned Capsule 1978," *Encyclopedia Astronautica*, http://www.astronautix.com/craft/chic1978.htm (accessed May 24, 2005).
Encyclopedia Astronautica. "Shuguang-1," *Encyclopedia Astronautica*, http://www.astronautica.com/craft/shuuang1.htm (accessed May 24, 2005).
Encyclopedia Astronautica. "ZY," *Encyclopedia Astronautica*, http://www.astronautix.com/craft/zy.htm (accessed June 2, 2002).
European Space Agency, "Crew Return Vehicle," http://www.esa.int/esaHS/ESARZS0VMOC_iss_0.html (accessed February 12, 2006).
Forden, Geoffrey. 2004. "The Military Capabilities and Implications of China's Indigenous Satellite-Based Navigation System," *Science and Global Security* 12, 219–250.
Ge, Lou. 2006. "China's Space Program," *Center for Security and International Studies* (April 3), http://www.csis.org/media/csis/events/060403_china_space.pdf (accessed April 11, 2006).
Go Taikonauts. "Launch Log," http://www.geocities.com/CapeCanaveral/Launchpad/1921/launch.htm (accessed May, 2005).
Grondine, E.P. "Chinese Manned Space Program: Behind Closed Doors," *Encyclopedia Astronautica*, http://www.astronautix.com/articles/chidoors.htm (accessed January 10, 2005).
Handberg, Roger. 1999. "Changing Parameters of Japanese Security Policy: The Advent of Military Space in the Post Cold War Environment." Paper presented at the Annual Meeting of the Southern Political Science Association, Savannah, November.
Handberg, Roger. 2003. "Dancing with the Elephants: Canadian Space Policy in Constant Transition," *Technology in Society* 25, 27–42.
Handberg, Roger. 1998. "The Fluidity of Presidential Policy Choice: The Space Station. The Russian Card and US Foreign Policy," *Technology in Society* 20, 421–439.
Handberg, Roger. 2004. "Outer Space as a Shared Frontier: Canada and the United States, Cooperation between Unequal Partners," *American Behavioral Scientist* 47 (June), 1263–1277.
Johnson-Freese, Joan. 2003. "China's Manned Space Program: Sun Tzu or Apollo Redux?" *Naval War College Review* 56 (Summer), 51–71.
Johnson-Freese, Joan. 2003. " 'Houston, We Have a Problem': China and the Race to Space," *Current History* (September), 259–265.
Johnson-Freese, Joan. 2005. "Testimony Before the US–China Economic and Security Review Commission: China Military Modernization and Cross-Strait Balance," September 15.
Kulacki, Gregory and David Wright. 2004. "A Military Intelligence Failure? The Case of the Parasite Satellite" (August 16), Union of Concerned Scientists, http://www.ucsusa.org/global_security/china/page.cfm?pageID=1479 (accessed May 21, 2006).
Lan, Chen. "Shenzhou History," http://www.geocities.com/CapeCanaveral/Launchpad/1921/story-7.htm?200517 (accessed March 17, 2005).

Launius, Roger D. 2004. "After Columbia: The Space Shuttle Program and the Crisis in Space Access," *Astropolitics* 2 (Autumn).

Lewis, John Wilson and Hua Di. 1992. "China's Ballistic Missile Programs: Technologies, Strategies, Goals (in Weapons Diffusion)," *International Security* 17(2) (Autumn), 5–40.

Li, Shenzhi. "When did Mao Zedong Decide to Lure Snakes Out of Their Holes?" *Works of Li Shenzhi*, http://members.lycos.co.uk/sixiang000/author/L/LiShenZhi/LiShenZhi008.txt (accessed June 9, 2006).

Li, Zhen and Roger Handberg. 2002. "The Central Dilemma of China's S&T Policy," *Bulletin of Science, Technology & Society* 22, 484–495.

Normile, Dennis and Ding Yimin. 2002. "Science Emerges from Shadows of China's Space Program," *Science* 296 (June 7), 1788–1791.

Robinson, Joan. 1968. "The Cultural Revolution in China," *International Affairs* (Royal Institute of International Affairs) 44(2) (April), 214–227.

Wade, Mark. 2006. "Shenzhou," *Encyclopedia Astronautica* (January 6), http://www.astronautrix.com/craft/shenzhou.htm (accessed February 15, 2006).

Yeh, Kungchia. 1993. "Foreign Trade, Capital Inflow, and Technology Transfer," in Michael Ying-Mao Kau and Susan H. Marsh (eds), *China in the Era of Deng Xiaoping: A Decade of Reform*. Armonk, NY: M.E. Sharpe.

Zhang, Hui. 2006. "Security in Space," *China Security*. Washington, DC: World Security Institute China Program.

Index

Aerospace Institute of China 115
Afghanistan 172
Alamaz military space staton 147
American-Soviet space race 4, 5, 27–28, 133, 153–154, 166
Annual Report to the Congress on the Military Power of the People's Republic of China (varying years) 116, 151, 158–159
Anti-Ballistic Missile (ABM) Treaty 20, 112, 159
Anti-satellite weapons (ASAT) 25, 74, 158–159
Apollo program 17, 22, 38, 41, 133, 139, 144, 168
"Apollo on Steroids" 139
APSat 107, 110
Ariane 4 106
Ariane 5 26, 106
Arianespace 38
Asia-Pacific Space Cooperation Organization (APSCO) 169
AsiaSat 110
Asymmetrical warfare 158
Atlas missile 59
Atlas 5 rocket 94, 140
Australia 38, 55
Automated Crew Transport (ATV) 147, 164

B-29 bomber 57, 60
B-36 bomber 57
B-47 bomber 57
B-52 bomber 57
Baikonur Cosmodrome 55
Ballistic missile defense (BMD) 20, 33, 113, 159
Bangladesh 169
Beidou system 45, 115–116
Beijing Aerospace Command and Control Center 140

Beijing Olympics 146
Beijing team 18, 79
Belgium 46
"Big Tiger" 131
Boeing 106
Brazil 1–3, 37, 102, 155
Brezhnev Doctrine 31–32
Brilliant Peebles 159
British Blue Streak 37
Budget capabilities 50–51
Bush, George H.W. 89
Bush, George W. 6, 22, 49, 112, 158, 159

Cal Tech 60
Canada 48, 55, 67, 128, 131, 147, 165, 170
Central Military Commission 32, 73
Central Party Committee 32, 62, 123–124
Challenger accident 128, 131
Chang'e lunar program 149
Chen Boda 77
Chen Shuibian 111
Chiang Kai-shek 29, 63, 64
China 47; agriculture 94–95; as symbol 129, 154–156; Civil War 29; conflict with Soviets 64–65, 81–82; economic modernization 124–126, 133–134, 153–154; economic and technological constraints 97–98; future in space 160–167; market 35; military weakness 59–60; national uniqueness 34, 39; political isolation 3, 4, 7, 9, 61; size 13–14, 15, 35, 98, 152–153; struggle to rebuild 28, 58; underdevelopment 3, 4, 58–59, 61–62, 68–69, 74–75, 91–92; vulnerability to U.S. attack 57–58, 62–63
Chinese Academy of Sciences 134–135

Chinese Academy of Space Technology 137, 153
China Aerospace Technology Group 121
China-Brazil Earth Resources Satellites (CBERS) 88, 97, 102–103
China emergence as space power 165
China-Japan rivalry 44, 113–114
China's National Defense, 2004 111
China National Space Administration 164
China-Russia space ties 139–140
ChinaSat 107, 110
China Satellite Communication Group Corporation 120
Chinese Communist Party 8, 29, 30, 31, 46, 58, 62, 66
Chinese foreign policy 9–10, 31–33, 57–59
Chinese launchers 11, 74
Chinese nationalism 133, 166, 167–168
Chinese nuclear program 18, 62–65
Chinese race to space 141, 142, 144–145
Chinese space program 1, 11; comparative analysis 14, 16–17, 58, 60, 63, 133, 138, 143, 144, 147–148, 149, 166, 167; cultural uniqueness 11; domestic implications 17; economic constraints 12–13, 15, 21, 23, 35, 43, 52, 67–68, 80–81; economic development 14, 30, 33, 46, 68, 90, 92–93, 96–97, 118–121, 163, 164; economic motivations 12–13, 119, 122, 127–129, 137–139; four eras 59; first era 59–70; fourth era 99–105; human space flight 15, 48, 73–74, 78–79, 117–118, 127–150; military component 11, 66, 92–93; military motivations 12, 45–46, 52–53, 57, 65, 66, 134; motivations 12–13, 16–17, 92–97; national prestige motivations 13, 24, 41, 117–118, 128–129, 147–148; opaque budget 23; second era 72–74, 70–82; second era decline 77–78; third era 84–99; relative normalcy 85, 101; partner 155–156; satellite programs 86–89; launch technologies 87–88; split with Soviet Union 15, 30–31, 41; successes 65; technological constraints 13; upheaval 2–3, 7, 10; uniqueness 36–37; weakness 20, 42–43
Chinese space program, 4 major issues 168–173; program stability 168–169; international cooperation 169–170; market involvement 171; military space 169–170

Chinese space station 148
Clinton, William 160
Cold War 1, 4, 14, 46, 111, 129, 142, 160
Columbia loss 47, 49, 53, 128, 130, 131
Commercialization of space 89
Commercial launch 104–111, 151–152; failures 108
Commercial launch failures (China) 104–106, 107–108, 151, 162–163
C^4I system (command, control, communications, computers and intelligence) 12, 116
Command economy 21
Communications satellites (comsats) 13
Corona spy satellite program 130–131
Crew Exploration Vehicle (CEV) 139, 145, 147
Crew Return Vehicle (CRV) 48, 147, 164
Critical pathway 165
Cuban Missile Crisis 18, 153
Cultural Revolution 7, 12, 115, 18, 31–32, 36, 59, 70–71, 72–74, 81–82, 90, 93, 96, 98, 101, 135, 152, 154
Czechoslovakia 32

Decline commercial launch market 107–110
Delta IV 26, 104
Deng Xiaoping 19, 31, 66, 67, 77, 80, 84, 85, 88, 89–91, 95–96, 99–101, 152
Democratic People's Republic of Korea 8
Denmark 46
Department of Postal Service 93
Developing states 7–8
Dismantling state controlled economy 91–92
DFH (Dong Fang Hong) communication satellites 66, 74, 84, 86, 88, 93, 99, 119
Domestic politics 38, 65–66, 70–71, 76–77, 80, 89
Dong Feng (DF)-1 missile 62, 63, 76
Dong Feng (DF)-2 missile 62, 64
Dong Feng (DF)-3 missile 33, 62, 64
Dong Feng (DF)-4 missile 64, 73, 76
Dong Feng (DF)-5 missile 64, 73, 76, 78
Dong Feng (DF)-6 missile 78
Dual-use 10–11, 44, 108, 111, 172

Early warning satellites 42
East Asia littoral 8
Economics and Economic competitiveness 21–24; continuous investment 21–22; high cost 24

Education reform 96–97
Eisenhower, Dwight 4–5, 24, 27–28, 40, 43, 57, 64
Eleventh National Party Congress 32
Embassy bombing, Belgrade 112
Environmental protection 121
Euraspace 108
Eureka Plan 100
Europa rocket 37–38, 55
European colonialism 29
European Launcher Development Organization (ELDO) 37
European Space Agency (ESA) 4, 38, 43–44, 45, 55, 92, 154, 147, 169
European Union (EU) 45
Europe 4, 7, 39, 43–44, 88–89, 104, 127, 134, 137, 155, 158, 159–160, 165, 170
Evolved Expendable Launch Vehicle (EELV) 22
Exceptionalism (China) 9
Exceptionalism (United States) 9
Explorer 1 (U.S.) 25
Explorer 1 (Kaituozhe-1) 104
Extra vehicular activity (EVA) 146

"Fear" as space program motivator 4
Feng Bao launcher 18
Fehng Huo-1 (FH-1) military comsat 116
Fei Junlong 145
Fifth Academy 78
Five Year Plan, First 30
Five Year Plan, Tenth 103, 104, 119, 121
Fixation on technologies 35–36
Force enhancement 41–42
Four Modernizations 32, 90–91, 95–96, 99–100, 137
France 28, 46, 67
French Indo China 9
French Coralie 37
FWS (Fan Hui Shi) recoverable satellites 84, 86, 86, 94, 103, 110, 105
FY Meteorological satellites 74, 86–87, 98–99

Galileo navigation satellite system 4, 45, 149, 156, 158, 165, 170
Gang of Four 32
Gagarin, Yuri 134, 146
Genesis spacecraft 131
Germany 38, 46, 50, 54–55, 72, 90
Glenn, John 146
Globalization effect 46
Gorbachev, Mikhail 128
Government controlled economy 171

GPS navigation satellite system 45, 158
Great Leap Forward 29, 66
Great Wall Industry Corporation 95, 104, 171
Griffin, Michael 164, 169
Gulf War (1991) 157

Harvey, Brian 12, 19, 128
Hermes spacecraft 4, 48, 137
Highly inclined elliptical orbit (HEO) 11
Hong Kong 26, 110
Hua Dong Computer Technology Research Institute 18
Hua Guofeng 89, 92
Hubble Space Telescope 49–50
Hu Jiantao 169
Human spaceflight 5–6, 47–49, 92, 102, 127–150; costs 131–133; Benefits 6; dangers 130–131; uniqueness 48
Hundred Flowers Campaign 30
Hu Yaobang 96

Incohin landing 30
India 2, 6, 13, 14, 24, 26, 34, 42, 46, 47, 72, 93, 113, 127–128, 159, 152, 155
Indo-china 30
Indonesia 113, 169
Inmarsat 172
Intelsat 107, 172
Intercontinental ballistic missile (ICBM) 58, 159
International Astronautical Federation 88, 138
International space regime 2, 153–156
International Space Station (ISS) 6, 48, 49–50, 54, 92, 131–132, 146–148, 165, 170
International Traffic in Arms Regulations (ITAR) 109–110
Iran 20, 33, 113, 169
Iraq 170
Iridium 106
Israel 48–49
Italy 38, 67

Japan 2, 4, 6, 7, 9, 19, 27, 29, 39, 42, 45, 48, 64, 67–68, 72, 92, 93, 127–128, 129, 131, 147, 154, 159, 170, 173; concern with China 44; H-IIa 44, 106
Jet Propulsion Lab 60
Jiaquan Satellite Launch Center 64–65, 79, 140
Jishu satellite series 18, 19, 72, 74, 79–80

Jianbing-3 photo reconnaissance satellite 115
Jiang Qing 67, 80
Jiang Zeimin 7, 18, 117, 123–124, 168
Johnson-Freese, Joan 11–12, 128, 133, 144
Johnson, Lyndon 133

Kazakhstan 55
Kennedy, John 17, 22, 42, 77
Kennedy Space Center 141
Khrushchev, Nikita 17–18, 25, 27, 30–31, 41, 58, 61, 77, 133
Kingdom of Tonga 55
Kliper spacecraft 128
Komarov, Vladimir 131
Korean War 5, 8, 9, 29–30, 57, 63, 64, 76, 158; Chinese intervention 8, 63
Kosovo 170
Kuomintang-CCP United Anti-Japanese Front 29
Kurs rendezvous system 139

Lack of trained personnel 61–62
Launch price controls 106–107, 123
Launch vehicles 25 failure rate 25–26, 37–38; 51; funding 26
Launius, Roger 6
Liability Convention 157
Li Qinqlong 139
Lin Biao 18, 19, 31, 32, 66, 67, 71, 72, 75–76, 77, 79, 135, 136, 152
Little Red Book 75
"little tigers" 131
Liu Shaoqi 31, 66, 67, 71, 75
Lockheed Martin 106
Long March, The 29
Long March rockets 26, 85, 95, 97, 104, 162; CZ-2E 85, 87, 103, 135, 138; CZ-2F 103–104, 138, 142, 143, 148; CZ-2G 148; CZ-3 84–85, 87, 103; CZ-4 85, 87, 103
Loss of cosmonauts 47
Low Earth orbit (LEO) 11, 50–51, 103
Lunar landing 8
Luo Ge 164

McNamara, Robert S. 20
Mao Zedong 5, 7, 9, 13, 17, 19, 29, 30, 31, 41, 58, 59, 61, 62, 65, 66, 68, 70–71, 74–76, 84, 87, 88, 89, 90, 96, 134, 141
Manchuria 29
Manned Orbital Lab (MOL) 148
Mao-Lin Coalition 74–75

Market economy 91–92, 95, 108, 125, 171, 172
Mars 6
Massive retaliation 4, 8–9
Mercury 7 141
Microgravity materials processing 94–95
Middle Altitude orbit (MAO) 11
Military modernization 96, 157–158
Military space 111–116
Ministry of Machinery Industry 67, 23, 78
Ministry of Post and Telecommunications 110
Mir space station
Mir 2 space station 147
Missile gap 31, 59
Missile Technology Control Regime (MCTR) 26, 42, 47, 155
Mongolia 169
Moon 6
Mutual Defense Assistance Agreement 63–64

National Aeronautical and Space Administration (NASA) 12, 61, 145, 147, 160, 164, 167, 168, 170
National air space 25
National Defense Science Committee 73
Nationalists 8, 29, 30, 63
National High Technology Research and Development Program 100–101
Nie Haishenq 145
Nigeria 155
Ninth National People's Congress 121–122
Nixon, Richard 9, 31, 96, 118
North Atlantic Treaty Organization (NATO) 30
North Korea 6, 8, 20, 29–30, 33, 44, 45, 90, 93, 113
Nuclear blackmail 63, 64
Nuclear stalemate 25
Nuclear weapons, China lack of 5; preemptive attack by U.S. 25, 31; testing 62

Opening to the West 85–86, 88–89, 91
Optus-B2 107
Orbital arc 10–11

Pakistan 127, 169
Parasitic satellite 116
Paper tiger 59–60
Patent applications 123
Peaceful uses of space 28

People's Liberation Army (PLA) 18, 29, 31, 32, 63, 71, 75–76, 135
People's War 76
Peru 169
Philippines 62, 64, 113
Pogo effect 130
Politics and space programs 10, 13, 38
Pontes, Marcos 48
Post Cold War 27, 88
Powers, Gary 57
"Prague Spring" 32
Pre-emptive strike 161–162
Program failure 51–52
Program intensity 52
Progress spacecraft 148, 164
Project 714 134–137; shut down 136
Project 863 100–101, 137; project 863-2 127; project 863-204 137; project 863-205 137
Project 921 137–145, 167
Proton rocket 105, 162

Qian Xuesen 73, 136
Qing (Manchu) emperor 28–29
Quadrennial Defense Review, 2001 112

R-1 rocket 60
Radiation belts (hazard) 87
Ramon, Ilan 48
Rationales for space initiatives 40; commercial 40; human spaceflight 40; military 40; scientific 40
RD-120 rocket engine 139
RD-180 rocket engine 140
RD-170 rocket engine 140
Reagan, Ronald 100, 128
Recoverable satellite program 74, 80–81
Red Guards 67, 71, 73
Remote sensing satellites 13–14, 23–24, 94; disaster monitoring 94
Republicans 108
Republic of China 9
Republic of Korea 8
Right of free passage 24–25
Rogue states 20
Russia 1, 11, 19, 22, 33, 41, 105, 128, 130–131, 139–140, 145, 147, 172, 165

Salyut space station 117, 131, 136, 147
Sanctuary 156–157
Saturn 1B 60
Saturn 5 60
Saudi Arabia 33
Science and technology development 121–122

Security dilemma 4, 19, 21
Seeding breeding satellites 119, 171
Self-Strengthening Movement 69
Seventh Ministry 62, 78
Shanghai faction 18–19, 77–78, 79
Shanghai Jiangnan shipyard 18
Shen Jian (or Divine Arrow) 138
Shenzhou-Soyuz comparison 139–140
Shenzhou (Divine Vessel) Program 139, 141–145, 146–147, 149, 164
Shenzhou 1 142
Shenzhou 2 142–143
Shenzhou 3 143
Shenzhou 4 143
Shenzhou 5 1, 3, 117, 143–144, 162, 166, 167
Shenzhou 6 145–146
Shenzhou 7 146–147, 167
Shenzhou 8 148
Shenzhou 9 148
Shenzhou 10 148
Sheppard, Alan 134
Shijian satellite 76, 77
Shuguanq-1 (Dawn) 74, 78, 135–137
Singapore 55
SJ (Shijian) science satellites 84, 86, 87
Skylab 137, 147
SOKOL space suit 139
Southeast Asia Treaty Organization (SEATO) 30
South Korea 90, 131
Soviet missile program 59
Soviet TKS system 148
Soviet Union 2, 4, 5, 7, 8, 9, 12, 16, 19, 22, 32, 37, 39, 81, 85, 96, 116, 130, 131, 133, 136, 156–157; Collapse 8, 88, 107, 165; split with China 15, 30–31, 57–58, 70, 76
Soyuz space craft 128, 137, 139–140, 147, 164
Space arms race 159–160
"Space firsts" 27
Spaceflight: economics 17, 20, 21, 24; military aspects 22, 27, 39; motivations 16–17, 39–40; political basis 35, 37; prestige 16, 18; symbol of power 16
Space Medical Institute 73–74, 135
Space Ministry 97
SpaceShipOne 54
Space shuttle 26, 48, 52, 105
Space Station Mir 128, 131, 147
Space technologies 23, 36; "sweetness" 26
Space tourism 54
Sputnik 1 3, 17, 24

Stalin, Joseph 29, 63
"Star Wars" 158
State Aeronautics Industry Commission 62
State Council 62, 100, 111
State Science and Technology Commission 138
Stokes, Mark 12–13
Strategic Air Command (SAC) 60
Strategic Defense Initiative (SDI) 100, 159
Sun Yat-sen 29

Taikonauts (or *yuhangyuans*) 3, 94, 130, 135–136, 141; political reliability 135
Taiwan (formerly Formosa) 8, 920, 32, 33, 63, 90, 96, 111–112, 113, 131, 173
Technology gap 134
Technology impact of 24–27; independence 34
Technology transfer 26–27
Telesdic 23, 106
Thailand 169
Third Plenum 89
Tiananmen Square 33
Titan missile 59
Tsien Hsue-shen 60
Typology, national space participation 49–56; crew spaceflight 54; economic development 49; hierarchy 53, 156; Level One 53–54; Level two 54–55; Level three 55; Level four 55–56; newly emergent 54; proactive space participation 49–56; space faring 53–54

United Kingdom 21, 37, 38, 50, 51, 72
United Nations (UN) 8, 29–30
UN Committee on Peaceful Uses of Outer Space 88
United States 1, 2, 4, 5, 7, 8, 11, 16, 22, 29, 37, 39, 47, 81; 105–107, 96, 116, 128, 130, 131, 133, 136, 148–149, 156–157, 165; hostility to China 8–9, 19–20, 30, 57, 63, 65; space race with China 166–167; unilateral decisions 48
U.S. Congress 108, 147
U.S. Department of Commerce 108
U.S. Department of Defense (DoD) 8, 160
U.S. Department of State 108

U.S. Office of Commercial Space Transportation 109–110
U.S. Quadrennial Defense Review Report (2001) 8
U.S. Quadrennial Defense Review Report (2006) 8

Vandenberg Air Force Base 116
Vanguard rocket 143
V-2 rockets 50
Vietnam 113
Vietnam War 31, 63, 65, 76
Vision for Space Exploration 6, 22, 128, 144, 149, 167, 169
Von Braun, Wernher 24, 58

Wade, Mark 139–140
Wang Bingzhang 78, 152
Wang Dayan 99
Wang Hongwens 79
Warsaw Pact 9
Waves of space program initiation 40–47; cost concerns 43; first wave 40–41; military space applications 45; resource issues 46; restrictions on technology 46–47; second wave 42–45; third wave 46–47
Weaponization of outer space 156–157, 158, 160
White Paper on Space Programs 115, 161–167
World Trade Organization 119
World War Two 2, 28, 54
Wu Jie 139

Xichang Launch Center 64–65
Xinjiang 86
Xizang 86
X-Prize 54

Yalu River 8
Yanping Chen 59, 96
Yang Liwei 1, 3, 144, 145, 146, 166
Yao Tongbin 73
Yuan Shikai 29
Yuan Wang 4 tracking ships 140

Zarya (Dawn) 136
Zhang Chungiao 79
Zhao Erlu 73
Zhao Kiuzhang 81
Zhao Ziyang 96
Zhou Enlai 32, 62, 73, 90, 152
ZY (Ziyan) resource satellites 88, 99, 115, 119